PERV

www.**transworldbooks**.co.uk

ALSO BY JESSE BERING

Why Is the Penis Shaped Like That?

The God Instinct

PERV

THE SEXUAL DEVIANT
IN ALL OF US

JESSE BERING

Doubleday

LONDON · TORONTO · SYDNEY · AUCKLAND · JOHANNESBURG

TRANSWORLD PUBLISHERS
61–63 Uxbridge Road, London W5 5SA
A Random House Group Company
www.transworldbooks.co.uk

First published in Great Britain
in 2013 by Doubleday
an imprint of Transworld Publishers

A CIP catalogue record for this book
is available from the British Library.

ISBNs 9780857520401 (hb)
9780857520418 (tpb)

Addresses for Random House Group Ltd companies outside the UK
can be found at: www.randomhouse.co.uk
The Random House Group Ltd Reg. No. 954009

The Random House Group Limited supports the Forest Stewardship
Council® (FSC®), the leading international forest-certification organisation.
Our books carrying the FSC label are printed on FSC®-certified paper.
FSC is the only forest-certification scheme supported by the leading
environmental organisations, including Greenpeace. Our paper procurement
policy can be found at www.randomhouse.co.uk/environment

Typeset in 11/14 pt New Caledonia
Printed and bound in Great Britain by
Clays Ltd, Bungay, Suffolk

2 4 6 8 10 9 7 5 3 1

MIX
Paper from
responsible sources
FSC® C016897

For you, you pervert, you

Rarely has man been more cruel against man than in the condemnation and punishment of those accused of the so-called sexual perversions. The penalties have included imprisonment, torture, the loss of life or limb, banishment, blackmail, social ostracism, the loss of social prestige, renunciation by friends and families, the loss of position in school or in business, severe penalties meted out for convictions of men serving in the armed forces, public condemnation by emotionally insecure and vindictive judges on the bench, and the torture endured by those who live in perpetual fear that their non-conformant sexual behavior will be exposed to public view. These are the penalties which have been imposed on and against persons who have done no damage to the property or physical bodies of others, but who have failed to adhere to the mandated custom. Such cruelties have not often been matched, except in religious and racial persecutions.

—Alfred Kinsey (1948)

CONTENTS

PREFACE

In 1985, when the AIDS epidemic and its concentrated scourge upon gay men were causing an unprecedented level of panic across America, I was an eminently underwhelming, overly sensitive ten-year-old boy living with my family in the leafy suburbs of Washington, D.C. This new disease—the "gay plague," as people were calling it—was suddenly the talk of our town. At a block cookout one summer evening, I sat near a group of men pontificating about "this AIDS thing." Looking back now, I don't think they even realized I was there; I was the sort of child who blended into tree bark and lawn ornaments. The men scratched their heads, threw back a few beers, did some entertaining imitations of outlandish drag queens, and then finally concurred that in all probability, in all seriousness, AIDS was just God's clever way of getting rid of the queers. (Like most of the men in my neighborhood, these comedians worked for the government, if I'm not mistaken.)

When I turned on the television back at home, I saw belligerent housewives and middle-school football coaches shouting antigay epithets at supporters of Ryan White, a gentle, eloquent adolescent with hemophilia who'd contracted HIV through a blood transfusion years earlier. The news footage showed his single mother wading patiently through an angry mob in her small

Indiana town to enroll her son in the public school. The grim death of an emaciated Rock Hudson that same year riveted people's attention, and with this attention came that terrible onslaught of jokes about fags and AIDS that saturated the talk in school cafeterias and on playgrounds, the residue of which can still be found in the bigoted banter of some chuckling adults to this day.

Now, by all appearances, I was an average boy; as I said, I didn't stand out in any way, which in this case means I wasn't your stereotypical "sissy." I certainly didn't play with dolls, anyway. Well, that's not entirely true. I adored my Superman doll. And what I adored about him most of all was stripping him nude and lying together naked under the covers. (Hugely disappointing, yet somehow each time the anticipation of finding more than a slick plastic crotch would build in my mind just the same.) But this AIDS fiasco made my burgeoning desires more salient to me than they probably otherwise would have been. The menacing ethos of those times, in which it was made abundantly clear to me that people like me were not welcome in this world, prematurely pushed a dim awareness of my own sexuality into my consciousness. What I didn't understand was that gay males were dropping like flies not because they—or rather *we*—were inherently bad and "disgusting" but because they'd engaged in a form of unprotected sex that made them especially vulnerable to the virus. I wasn't an epidemiologist. I was a fifth grader. I didn't even know what sex *was*.

To my mind, gays were simply being struck down one by one by a mad God, just as I'd heard those men saying at the cookout. So my days, I figured, must be numbered too. When would *I* start showing those telltale sores on my face, or perhaps the grayish pallor, the strained breathing, the zombielike gait of the other "positive" ones that I kept seeing on television and in the newspapers? One day I stood before the mirror and lifted up my shirt only to find a loom of prepubescent ribs that served to convince me I had indeed started wasting away from this unholy affliction.

In reality, I was just extra scrawny. But my flawed religious interpretation of what was happening is all the more revealing of the caustic moralism of the times given that my family was by no means religious.

I couldn't share my crippling anxiety with my perfectly reasonable parents. That would mean the unthinkable risk of outing myself as one of these social pariahs that everyone was talking about. My fears intensified when I realized that concerted efforts to suss us out from the "normal" people were already well under way. From scattered threads of gossip and the occasional sound bites, I managed to piece together that the best way to detect our *essential evil*, to reveal what God alone already knew, was to analyze our blood for evidence of some kind of gay particle. It was only a matter of time before a stern-faced scientist would hold a test tube up to the light and exhibit before a hushed gathering of his peers how my hidden nature danced and mingled in all its monstrous opaqueness against the pure rays of the sun. In the meantime, I stuck my head out the car window and screamed "Faggot!" at my older brother—who was then just as he is now, about as straight as straight gets—while he was playing in the street, just to throw off the undercover witch-hunters in the neighborhood. As we all know perfectly well, a person who shouts homophobic slurs can't *possibly* be gay.

As my annual doctor's visit approached ominously on the calendar, my measured apprehension (too strident a protest would only give me away) failed to register to my parents as anything more than run-of-the-mill cowardice. The irony is that by the time I dragged my feet into the pediatrician's office and the needle was plucked from my arm after a routine blood draw, all those months of stress boiled over into a very nonimaginary illness. On seeing my liquidized evil lapping forebodingly in a vial in the nurse's white-gloved hands, I became so instantly sick over my now inescapable fate that I grew faint and then threw up all over the phlebotomist's chair. Imagine my relief when the absentminded doctor—probably, I thought, just distracted by all the commotion—miraculously missed my dark secret and didn't have to break the unspeakable news to my parents.

It would be a decade before I dared to come out to them, and by then they'd divorced. I decided to break the news to my mom first. She was a warm person with a good sense of humor that was tempered (sadly, too often) by a tragedian air to her personality. I'd no doubt she'd still love me when all was said and done, but I also knew she could be willfully naive about subjects that frightened her or made her uncomfortable. Sex was a big one. I never heard her utter a hateful word about gay people, but neither can I recall her ever saying anything positive. Homosexuality was just a nonissue in our house. Or so she thought.

In the kitchen one evening, I blurted out that I had something I needed to tell her. I sat at the table fiddling nervously with the edges of the newspaper. "*What*?" she said just as nervously. "Jesse, what *is* it?" She went on, prodding me. "I'm gay," I said. It was the first time I'd ever said it aloud, and I felt my ears ring at the sound of it. "Oh, come on," she said through a widening grin, figuring I must be playing a joke on her. "*No*. You're kidding. Aren't you?" "No," I said. "I really am, I mean, I really am gay."

I'd long prided myself on my deceptive use of language. A strategically placed hesitation, a subtle omission of fact, a carefully inserted sigh, a sibilant hiss that lasts but a second, the intonation of a vowel to fill it with a mirage of meaning, these and more were all in my arsenal of verbal legerdemain. It had kept me safe all this time. Just look: I'd even tricked the woman in whose uterus my brain first began wiring itself in a way that would lead directly, some twenty years later, to this excruciatingly awkward moment. My solitary and bookish ways as a little boy, the fabricated girlfriends, sublimating myself with schoolwork that first year of college, the meticulously kept collection of *Men's Fitness* magazines piled high in my closet throughout high school (I can't believe she didn't catch on with *that* one), it all clicked for her in that single snap of time. *She had a gay son.* I watched her breathe her last gasp of maternal denial. This was replaced, for a while, by stoic caregiving: she wasn't happy about my revelation; it was more a grin-and-bear-it type of situation. Years later, she confided in me that she'd had nightmares for the next six months featuring me in women's clothing and makeup, prancing around with strange men. I could only assure her that cross-dressing was one thing she definitely didn't need to worry about with me; my fashion sense was so abysmal, I reminded her, that I barely knew how to dress myself as a man, let alone pull off female couture. (Or perhaps that's exactly what she was worried about, now that I think about it.)

In any event, she got over it. So much so that by the time she succumbed to cancer only five years after this overdue tête-à-tête, I think the fact that her youngest son was gay had become a vague source of pride for her. I'd forcibly peeled it apart like a reluctant flower in the kitchen that day, yet ultimately my confession opened up her mind to a new way of thinking. Her nice but mostly uneventful suburban life was cut too short, but in her remaining years she quite literally fought to the death for me. She left this world on the side of reason, even if that meant exchanging words with her own mother, my cloistered eighty-two-year-old grandma, who was under an even more unshakable impression that gay men were transvestites. Mom, I'm glad to say, ultimately straightened Grandma out on that one.

When I struck up the courage to tell my father, an affable glue salesman with a penchant for quoting bisexual poets, I could only wonder why I hadn't told him years earlier. Consistent with his it-is-what-it-is philosophy on life, he shrugged, asked how I was doing in school, and told me he was sure I'd meet a nice boy soon enough.

It's still far from being ideal for gay youth, but there's genuine reason for them to be optimistic about their future. Much more than I was, anyway. The HIV panic has subsided, and we now know much more about how the virus is transmitted and how to prevent its spread. Although AIDS remains a crisis among certain communities (gay or otherwise), HIV is no longer a death sentence. In the United States and many other countries, gays and lesbians have also found increasing acceptance, with bigots now being vehemently called out as such by influential public figures. The toxic milieu of the mid-1980s that was personified by the heavy-metal singer Sebastian Bach wearing a T-shirt on national TV reading "AIDS Kills Fags Dead" is long gone. And good riddance. Today there are gay youth advocacy initiatives like the "It Gets Better" campaign, which was launched in 2010 by the advice columnist Dan Savage and his husband, Terry Miller, in response to an alarming rash of gay teen suicides.

I've benefited from this sea change as well. In 2006, after a stint as a psychology professor in Arkansas (of all places), I immigrated with my partner to Northern Ireland (again, of all places) for an academic appointment in Belfast. Soon after we arrived there, Juan and I entered into a "civil partnership"— turns out my father was right about me meeting a nice boy—a legal arrangement that granted us the rights of any straight married couple in the United Kingdom. When one considers how this particular region is synonymous with conservative religious beliefs (think of the Troubles and that interminable clash between Protestants and Catholics), the formal recognition of a gay couple as being legally equivalent to a married man and woman is a remarkable social accomplishment (even if the clerk in Belfast City Hall *did* complete our paperwork through a begrudging series of sighs and warned us of the Leviticus-riddled picket signs in the courtyard). Just like a thrice-divorced man married to the hooker he met at a fish-and-chips shop the night before, I was in a romance sealed with an ironclad decree approved by the British Crown.

Upon our return to America half a decade later, full-fledged marriage equality had already become a legislated reality in multiple U.S. states. In the mail just today, in fact, I received an invitation to my lesbian cousin's upcoming wedding in Connecticut. I'd like to think that even our squeamish late grandmother would have embraced her queer grandchildren by now. Once the shock wore off, I'm sure she'd find some humor in the fact that my gay Mexican partner makes me matzo ball soup using her favorite recipe (translated from the Yiddish) and that her lesbian granddaughter's fiancée is currently "knocked up" with a child conceived by artificial insemination.

~

At thirty-seven, I've already seen enormous change in my lifetime. It's all been for the better. Yet something has made me feel increasingly uncomfortable—or perhaps "guilty" is a better word. In the rush to redress the historical prejudice against gay people, we're missing a key opportunity as a society to critically examine our uneasy relationship with sexual diversity as a whole. We should certainly celebrate the fact that the lives of those who fit the LGBT (lesbian/gay/bisexual/transgender) label are improving, but we also shouldn't lose sight of the fact that those who can't be squeezed so neatly into this box are still being ostracized, mocked, and humiliated for having sexual natures that, if we're being honest, are just as unalterable. Apologies should be applied only to the things we've done wrong, not for who we unalterably are. I have a few scars that never healed properly from those ancient days when I was a terrified kid growing up gay in a climate of such intense scorn. This book, you might say, is my retaliation by reason. But I've come to realize that it's no longer gays and lesbians who need the most help. They could always use more, and I'm certainly here to weigh in on their behalf in the pages ahead and in real life, but today children like I once was have legions of fearless and vocal advocates. By contrast, many of these others—these "erotic outliers"—still live lives in constant fear for no reason other than *being*. And in fact there are many people, of all ages, who fit that bill.

What you're going to discover along the way is that you have a lot more in common with the average pervert than you may be aware. I'll be sharing with you a blossoming new science of human sexuality, one that's revealing how "sexual deviancy" is in fact far less deviant than most of us assume. Yet as we focus in on these glistening new findings of what secretly turns us on and off, it will also become increasingly apparent to you that the full suite of our carnal tastes is as unique to us as our fingerprints. When we combine this new science with forgotten old case studies showcasing some of the most bizarre forms of human sexuality, you'll catch a glimpse of the nearly infinite range of erotic possibilities. Finally, you'll come to understand why our best hope of solving some of the most troubling problems of our age hinges entirely on the *amoral* study of sex.

It's virgin territory indeed, but there's no time like the present, so let's dig in and penetrate this fuzzy black hole, shall we? I can't promise you an orgasm at the end of our adventure. But I *can* promise you a better understanding of why you get the ones you do.

PERV

ONE

WE'RE ALL PERVERTS

Gnothi seauton [Know thyself]
—Inscription outside the Temple of Apollo at Delphi

You are a sexual deviant. A pervert, through and through.

Now, now, don't get so defensive. Allow me to explain. Imagine if some all-powerful arm of the government existed solely to document every sexual response of every private citizen. From the most tempestuous orgasmic excesses, to the slightest twinges of genitalia, to unseen hormonal cascades and sub-cranial machinations, not a thing is missed. Filed under your name in this fictional scientific universe would be your very own scandalous dossier, intricate and exhaustive in its every embarrassing measurement of your self-lubricating loins. What's more, the records in this nightmarish society extend all the way back to your adolescence, to the days when your desires first began to simmer and boil. I'd be willing to bet that buried somewhere in this relentless biography of yours is an undeniable fact of your sex life that would hobble you instantaneously with shame should the wrong individual ever find out about it.

To break the ice, I'll go first. And how I wish one of my first sexual experiences were as charming as inserting my phallus into a warm apple pie. Instead, it involves pleasuring myself to an image from my father's old anthropology textbook. This isn't even as admirable as those puerile stories about a teenage boy masturbating to some *National Geographic*–like spread of exotic naked villagers breast-feeding or shooting blow darts in the Amazon. No, it wasn't anything like that. For me, the briefest of heavens could instead be found in an enormous and hairy representative of the species *Homo neanderthalensis*. I can still see the lifelike rendering now. The Neanderthal was shown crouching down, pink gonads dangling teasingly between muscular apish thighs, while with all his cognitive might this handsome, grunting beast tried desperately to light a fire in a cobbled pit to warm his equally hirsute family (what looked to be a perplexed woman from whose furry breasts a baby feverishly suckled). The Neanderthal was in fact too brutish for my tastes, but in those pre-Internet days he was the only naked man I had at my fingertips. Well, the only naked hominid, anyway. One must work with the material one has.

So there, I said it. In my adolescence, I derived an intense orgasm (or twenty) from fantasizing about a member of another species. (In my defense, it *was* a closely related species.) You may have to rack your brains for some similarly indecent memory, or then again, maybe all you need to do is roll over in bed this morning to remind yourself of the hairy specimen of a creature that *you* brought home last night. Either way, chances are there's something gossip-worthy in your own sexual past. Maybe it's not quite as odd as mine. But I'm sure it's suitably humbling for present purposes. What makes us all the same is our having had certain private moments that could get us blackmailed.

Granted, most of us will never share our own lurid tidbits about our most unusual masturbatory mental aids or the fact that there's a distinct possibility we had the tongue of a Sasquatch in our nether regions last night (or ours in its). What usually gets out is only what we want others to know. That's perfectly understandable. We have our reputations to consider. I might never be allowed again into my local museum for fear I'll debase one of the caveman mannequins, for instance. The problem with zipping up on our dirtiest little secrets, however, is that others are doing exactly the same thing, and this means that the story of human sexuality that we've come to believe is true is, in reality, a lie. What's more, it's a very dangerous lie, because it convinces us that we're all alone in the world as "perverts" (and hence immoral monsters) should we ever deviate in some way from this falsely conceived pattern of the normal. A lot of human nature has escaped rational understanding because we've been unwilling to be completely honest about what *really* turns us on and off—or at least what's managed to do the trick for us before. We cling to facades. We know one another only partially. Much of what lies ahead, therefore, concerns what you don't want the rest of the world to know about your sexuality. But relax, that will be our little secret.

Again, however, I'd urge you to come clean in the confession booth of your own mind. And really, just a small unburdening of your erotic conscience will do for now. Reach far, far into the abyss of your wettest of dreams. Or perhaps it was only a fleeting, long-forgotten secretion, a lingering gaze misplaced, a furtive whiff of an object redolent with someone you once craved, a wayward click of the mouse, a hypothalamic effervescence that made you tingle down below. Nevertheless, even if you settle on one of these relatively minor examples, each embodies a corporeal reality specific to *you* . . . a "shocking," incontrovertible deed of physiology or an outright commission of lust that you've never shared with a single person, maybe not even yourself until now.

Whatever it is, once it's laid bare for all the world to see in your declassified government report, a faultless testimony in inerasable ink, this unique venereal data point will undoubtedly register in the consciousness of someone, somewhere out there as evidence of your sexual deviance, or perhaps even your criminality. Just look around you or think of all the people you know. In the unforgiving lair of another's critical eyes, you have now been transformed irreversibly into a filthy, loathsome pervert. And *that's* the feeling, this fetid social emotion of shame, that I want you to keep in the back of your mind as you read this book. We're going to get to the bottom of where it comes from, and we're going to do our best to smother it with reason in our efforts to stop it from hurting you and others in the future.

This feeling doesn't just make you a guilty pervert; more important, it makes you a human being. Blue-haired grandmothers, somnambulant schoolteachers, meticulous bankers, and scowling librarians, they've felt it too, just like you. We tend not to think of others as sexual entities unless they've aroused us somehow, but with the exception of those people spared by certain chromosomal disorders, we're all innately lewd organisms. That's easy to grasp in some abstract sense. But try putting it into practice. The next time you're at the grocery store and the moribund cashier with the underbite and the debilitating bosom sweeps your bananas across the scanner, think of precisely where those uncommonly large hands have been. How many men or women—*including* her—have those seemingly asexual appendages brought ineffable bliss? This isn't an exercise in the grotesque; it's a reminder of your animal humanity. A concupiscent beast has roamed under all skins . . . even that of the grumpy checkout lady.

Yet the best-kept secret is even bigger than this unspoken universality. It's this: exploring the outer recesses of desire by using the tools of science is a pinnacle human achievement. It's not easy, but digging into the darkest corners of our sexual nature (that is to say, our "perversions") can expose what keeps us from making real moral progress whenever the issues of equality and sexual diversity arise. With each defensive layer we remove, the rats therein will flee at the daylight falling at their feet, and the opportunity to eradicate such a pestilence of fear and ignorance makes the excavation of our species's lascivious soul worth our getting a little dirty along the way.

We're not the first to use the grimier realities of human sexuality to grease our way into some deeper truths. They may not have been scientists, but many artists and writers have touched on related psychological processes that were insightful and even foretold future research directions. In his 1956 play, *The Balcony*, for example, the French playwright Jean Genet showed how people who are inebriated by desire experience cognitive distortions motivating them to engage in behaviors that in a less aroused state of mind they'd perceive as obscene. Genet's story revolves around the daily affairs of a busy brothel in a town on the brink of war. Run by an astute madam named Irma, the whorehouse is a sanctuary in which high-profile local officials are free to drain away their carnal excess. Once they've done so, they can get on with the business of being "normal" and respectable public figures defending the town from the enemy. Irma's house of illusions has come to serve some colorful patrons, including the town judge, who feigns to "punish" a naughty prostitute, a bishop who pretends to "absolve the sins" of a demure penitent, and a general who enjoys riding his favorite (human) horse. "When it's over, their minds are clear," Irma reflects after these men visit her establishment. "I can tell from their eyes. Suddenly they understand mathematics. They love their children and their country." The lustful human brain, Genet understood in a way that contemporary scientists are just now starting to fully grasp by using controlled studies in laboratory settings, is simply not of the same world as that of its sober counterpart.

One point I'd like to make crystal clear at the outset of our journey is that *understanding* is not the same as *condoning*. Our sympathies can take us only so far, and entering other minds isn't pleasant when it comes to certain categories of sex offenders. Furthermore, it's one thing to wax theoretical about sexual deviance, but another altogether to be the victim of sex abuse in real life or to know that someone we love, especially a child, has been harmed. Yet while it's a common refrain to liken the most violent sex offenders to animals, whether we like it or not, even the worst of them are resoundingly human. As unsettling as it can sometimes be to lean in for a closer look, their lives can offer us valuable lessons about what can go wrong in the development of a person's sexual identity and decision making. "I consider nothing that is human alien to me," said the Roman philosopher Terence. I feel the same way. And Terence's credo is one I intend to adhere to closely when it comes to some of the characters we'll be meeting along the way.

I'll do my best, anyway. For while there's no doubt that the most terrible rapists, child molesters, and other more banal classes of sex offenders were around in his day, Terence didn't know of the hundreds of extravagant "paraphilias" (or sexual orientations toward people or things that most of us wouldn't consider to be particularly erotic) that scientists would eventually discover when he confidently uttered those words more than two thousand years ago. Even he might have had trouble finding common ground with, say, "teratophiles," those attracted to the congenitally deformed, or "autoplushophiles," who enjoy masturbating to their own image as cartoonlike stuffed animals.

Understanding the etymology of the word "pervert," oddly enough, can help us to frame many of the challenging issues to come. Perverts weren't always the libidinous bogeymen we know and loathe today. Yes, sexual mores have shifted dramatically over the course of history and across societies, but the very word "pervert" once literally meant something else entirely than what it does now. For example, it wouldn't have helped his case, but the peculiar discovery that some peasant during the reign of Charles II used conch shells for anal gratification or inhaled a stolen batch of ladies' corsets while touching himself in the town square would have been merely coincidental to any accusations of his being perverted. Terms of the day such as "skellum" (scoundrel) or reference to his "mundungus" (smelly entrails) might have applied, but calling this man a "pervert" for his peccadilloes would have made little sense at the time.

Linguistically, the sexual connotation feels so natural. The very ring of it—*purrrvert*—is at once melodious and cloying, producing a noticeable snarl on the speaker's face as the image of a lecherous child molester, a trench-coated flasher in a park, a drooling pornographer, or perhaps a serial rapist pops into his or her head. Yet as Shakespeare might remind us, a pervert by any other name would smell as foul.

For the longest time, in fact, to be a pervert wasn't to be a sex deviant; it was to be an atheist. In 1656, the British lexicographer Thomas Blount included the following entry for the verb "pervert" in his *Glossographia* (a book also known by the more cumbersome title *A Dictionary Interpreting the Hard Words of Whatsoever Language Now Used in Our Refined English Tongue*): "to turn upside down, to debauch, or seduce." All of those activities occur in your typical suburban bedroom today. But it's only by dint of our post-Victorian minds that we perceive these types of naughty winks in the definition of a term floating around the old English countryside. In Blount's time, and for several hundred years after he was dead and buried, a pervert was simply a headstrong apostate who had turned his or her back on the draconian morality of the medieval Church, thereby "seducing" others into a godless lifestyle.

Actually, even long before Blount officially introduced perverts to the refined English-speaking world in all their heathen fury, an earlier form of the word appeared in the Catholic mystic Boethius's *Consolation of Philosophy* in the year 524.* Like

---

*Boethius's *Consolations* went largely unnoticed until Chaucer translated the mystic's treatise in the fourteenth century. Thomas Blount's seventeenth-century definition of "pervert" likely originates in turn from Chaucer's earlier translation of that 524 text.

Blount's derivation, the mystic's *pervertere* was a bland "turning away from what is right." Given the context of Christian divinity in which Boethius's treatise was written, it's clear that "against what is right" meant much the same then as it does for God-fearing people today, which is to say, against what is biblical.

So if we applied this original definition to the present icono-clastic world of science, one of the world's most recognizable perverts would be the famous evolutionary biologist Richard Dawkins. As the author of *The God Delusion* and an active pros-elytizer of atheism, Dawkins encourages his fellow rationalists to "turn away from" canonical religious teachings. (I've penned my own scientific atheistic screed, so I'm not casting stones. I'm proudly in possession of a perverted nature that fits both the ar-chaic use of the term, due to my atheism, and its more recent pejorative use, due to my homosexuality.)

Only at the tail end of the nineteenth century did the word "pervert" first leap from the histrionic sermons of fiery preachers into the heady, clinical discourses of stuffy European sexologists like the ones you'll be introduced to soon. And it was a long time after that still before "pervert"—or "perv" if we're being casual—became slang for describing the creepy, bespectacled guy up the road who likes to watch the schoolgirls milling about the bus stop in their miniskirts while he sips tea on his front porch.

This semantic migration of perverts, from the church pews to the psychiatric clinic to the online comments section of news stories about sex offenders, hasn't occurred without the clatter-ing bones of medieval religious morality dragging behind. Notice the suffix -*vert* means, generally, "to turn": hence "con*vert*" (to turn to another), "re*vert*" (to turn to a previous state), "in-*vert*" (to turn inside out), "per*vert*" (to turn away from the right course), and so on. But of all these related words, "pervert" alone has that devilishly malicious core—"a distinctive quality of obsti-nacy," notes the psychoanalyst Jon Jureidini, "petulance, peevish-ness . . . self-willed in a way that distinguishes it from more 'innocent' deviations." A judge accusing someone of "perverting the course of justice" is referring to a deliberate effort to thwart moral fairness. Similarly, with the modern noun form of "pervert" being synonymous with "sex deviant," the presumption is that he (or she) is a deviant by his own malicious design. That is, he is presumed to have willfully chosen to be sexually aberrant in spite of such a decision being morally wrong.

It's striking how such an emotionally loaded word, one that undergoes almost no change at all for the first thousand years of its use in the English language, can almost overnight come to mean something so very different, eclipsing its original intent in its entirety. So how, exactly, did this word "pervert" go from being a perennial reference to the "immoral religious heretic" to referring to the "immoral sexual deviant"?

The answer to this riddle can be found in the work of the Victorian-era scholar Havelock Ellis of South London, who is credited with popularizing the term in describing patients with atypical sexual desires back in 1897. Although earlier scholars, including the famous Austro-German psychiatrist Richard von Krafft-Ebing, regarded by many as the father of studies in deviant sexuality, preceded Ellis in sexualizing the term, Ellis's accessible writing in the English language found a wider general audience and ultimately led to the term being solidified this way in the common vernacular. The provenance of the term in Ellis's work is still a little hard to follow, because he initially uses "perverts" and "perversions" in the sense of sexual deviancy in the pages of a book confusingly titled *Sexual Inversion*. Coauthored by the gay literary critic John Addington Symonds and published posthumously, the book was a landmark treatise on the psychological basis of homosexuality. "Sexual inversion," in their view, reflected homosexuality as being a sort of inside-out form of the standard erotic pattern of heterosexual attraction. That part is easy enough to

understand. Where Ellis and Symonds's language gets tricky, however, is in their broader use of "sexual perversions" to refer to socially prohibited sexual behaviors, of which "sexual inversion" was
just one. (Other classic types of perversions included polygamy,
bestiality, and prostitution.) The authors adopted this religious
language not because they personally believed homosexuality to
be abnormal and therefore wrong (quite the opposite, since their
naturalistic approach was among the first to identify such behaviors in other animals) but only to note how it was so salient among
the categories of sexuality frequently depicted as "against what
is right" or sinful.* Also Symonds, keep in mind, was an out and
proud gay man. The word was merely an observation about how
homosexuals (or "inverts") were regarded by most of society.

Interestingly enough, the scientist of the pair, and the one
usually credited with christening gays and lesbians as sex "perverts," had his own unique predilections. Havelock Ellis's "uro-
philia," which is a strong sexual attraction to urine (or to people
who are in the process of urinating), is documented in his various notes and letters. In correspondence with a close female ac-
quaintance, Ellis chided the woman for forgetting her purse at
his house, adding saucily, "I've no objection to your leaving *liquid gold* behind." He gave in to these desires openly and even
fancied himself a connoisseur of *pisseuses*, writing in his autobiography: "I may be regarded as a pioneer in the recognition of
the beauty of the natural act in women when carried out in the
erect attitude." In his later years, this "divine stream," as he
called it, proved the cure for Ellis's long-standing impotence.
The image of an upright, urinating woman was really the only
thing that could turn him on. And he was entirely unashamed of

---

*Ellis was also a vehement supporter of eugenics and even held court for a while as the
president of the Galton Institute, an organization that sought to improve the fitness (or
reproductive quality) of our species's genetic stock through carefully regimented human breeding. The heritability of kinkiness didn't seem to be of any special concern
to him.

this sexual quirk: "It was never to me vulgar, but, rather, an ideal interest, a part of the yet unrecognised loveliness of the world." On attempting to analyze his own case (he was a sexologist, after all), Ellis concluded, "[It's] not extremely uncommon . . . it has been noted of men of high intellectual distinction."* He was also convinced that men with high-pitched voices were generally more intelligent than baritones. That Ellis himself was a rare high tenor might have had something to do with that curious hypothesis as well.

Ellis was among a handful of pioneering sexologists in the late nineteenth and early twentieth centuries who'd set out to tease apart the complicated strands of human sexuality. Other scholars, such as Krafft-Ebing, as well as the German psychiatrist Wilhelm Stekel and, of course, the most famous psychoanalyst of all, Sigmund Freud, were similarly committed to this newly objective, amoral empirical approach to studying sexual deviance. Their writings may seem tainted with bias to us today, and in fact they are, but they also display a genuine concern for those who found themselves, through no doing or choice of their own, being aroused in ways that posed serious problems for them under the social conditions in which they lived.

It's worth bearing in mind, for instance, that Ellis and Symonds's *Sexual Inversion* was written on the heels of Oscar Wilde's sensationalized 1895 gross indecency trials, in which (among other things) that great Dubliner wit was publicly accused

---

*Believing that sexual fetishes get their start in childhood experiences, Ellis recalled his mother playfully thrusting a younger sibling's wet diaper in his face when he was nine years old. Several years later, while strolling with the pubescent Havelock through the grounds of the zoo gardens, Mrs. Ellis lifted her skirt and squatted to relieve herself behind some bushes. Havelock remembered how the sound of her urine stream meeting the composted earth had titillated him as a twelve-year-old boy. Before long, he'd be scientifically measuring the distance and trajectory of his schoolfellows' pee—"my own vesical energy being below the average," he noted with a characteristically morbid self-analysis. His curiosity culminated in an empirical study: "The Bladder as a Dynamometer," *American Journal of Dermatology* 6 (May 1902).

of cavorting with a fleet of boys and men in a series of racy homosexual affairs. Taking the stand at London's Old Bailey courthouse, where the father of his petulant young British lover, Lord Alfred Douglas, had brought charges against him, Wilde famously referred to homosexuality as "the love that dare not speak its name." The jury sentenced him to two years of hard labor for the crime of sodomy. (Incidentally, although consensual anal sex is no longer a crime in the United Kingdom, the fact that forcible anal penetration, among other acts, is still officially called "sodomy"—as in Sodom and Gomorrah—throughout the industrialized world even today shows just how deeply an antiquated religious morality is embedded and tangled up in our modern sex crime laws.)

What often gets overlooked in Wilde's account is the fact that "the love that dare not speak its name" referred to a specific type of homosexual relationship. Sexologists today would label Wilde's well-known affinities as evidence of his "ephebophilia" (attraction to teens or adolescents).* Wilde's intent in the phrase being especially applicable to courtships between men and teenage boys is clear when one reads his full elaboration on the stand, where he goes on to describe this unspeakable love:

> As there was between David and Jonathan, such as Plato made the very basis of his philosophy, and such as you find in the sonnets of Shakespeare. It is that deep, spiritual affection that is as pure as it is perfect. It dictates and pervades great works of art like those of Michelangelo . . . It is beautiful, it is fine, it is the noblest form of affection. There is nothing unnatural about it. It is intellectual, and it repeatedly exists between an elder and a younger man,

*See, for example, the French novelist André Gide's firsthand account of his traveling to Algiers with Wilde to procure adolescent boys for sex not long before the trial: *If It Die: An Autobiography*, trans. Dorothy Bussy (New York: Modern Library, 1935).

when the elder man has intellect, and the younger man has all the joy, hope and glamour of life before him. That it should be so, the world does not understand. The world mocks at it and sometimes puts one in the pillory for it.

Wilde's description of such a mutually beneficial, intergenerational romance is ironic today, because "the love that dare not speak its name" is now more unutterable than ever. The modern ephebophilic heirs of Wilde, Plato, and Michelangelo are not only mocked and pilloried but branded erroneously, as we'll see later, as "pedophiles."

Much like Wilde facing his detractors, the early sexologists found themselves confronted by angry purists who feared that their novel scientific endeavors would open the door to the collapse of cherished institutions such as marriage, religion, and "the family." Anxieties over such a "slippery slope effect" have been around for a very long time, and in the eyes of these moralists an objective approach to sexuality threatened all that was good and holy. Conservative scholars saw any neutral evaluation of sex deviants as a dangerous stirring of the pot, legitimizing wicked things as "natural" variants of behavior and leading "normal" people into embracing the unethical lifestyles of the degenerate.* Merely giving horrific tendencies such as same-sex desires their own proper scientific names made them that much more real to these moralists, and therefore that much more threatening. To them, this was the reification of sexual evil. In a scathing

---

*Getting wind of Ellis and Symonds's soon-to-be-published book, the Manhattan psychiatrist Allan McLane Hamilton put out his "Civil Responsibility of Sexual Perverts," *American Journal of Psychiatry* 52 (1896). Whereas Ellis had mostly a shoulder-shrugging philosophy about homosexuality, Hamilton felt that it was so corrupting a disease that anyone found in a committed gay or lesbian relationship should be separated by force. "I hold that under such circumstances not only may the aid of habeas corpus be implored for the purpose of effecting a separation, but that in aggravated instances the physician should, in manner specified, bring the matter before the attention of a committing judge" (511).

review of *Sexual Inversion*, for instance, a psychiatrist at the Boston Insane Hospital named William Noyes chastised the authors for "adding three hundred more pages to a literature already too flourishing . . . Apart from its influence on the perverts themselves no healthy person can read this literature without a lower opinion of human nature, and this result in itself should bid any writer pause."

Looking back now, it becomes evident that Ellis and Symonds's careful distinction between homosexual *behavior* and homosexual *orientation* was an important step in the history of gay rights. It may seem like common sense today, but for the first time ever homosexuality was being widely and formally conceptualized as a psychosexual trait (or orientation), not just something that one "did" with members of the same sex.* This watershed development in psychiatrists' way of thinking about homosexuality had long-lasting positive and negative implications for gays and lesbians. On the positive side, homosexuals were no longer perceived (at least by experts) as fallen people who were simply so immoral and licentious that they'd even resort to doing *that*; instead, they were seen as having a psychological "nature" that made them "naturally" attracted to the same sex rather than to the opposite sex.

On the negative side, this newly recognized nature was also regarded as inherently abnormal or flawed. With their inverted pattern of attraction, homosexuals became perverts in essence, not just louses dabbling in transgressive sex. Whether or not they ever *had* homosexual sex, such people were now one of "them."

*In *The History of Sexuality, Vol. 1* (New York: Random House, 1978), the philosopher Michel Foucault traces the origins of homosexuality as biological essence rather than action to a rather obscure 1870 paper by the German neurologist Carl Westphal. Westphal had described several patients with "contrary sexual feelings" that today we'd recognize clearly as being gay men and lesbians. It's splitting hairs really, but in my reading of the historical literature, it was only with the extensive treatments of the subject by later sexologists such as Krafft-Ebing and Ellis (combined with the further reach of their popular books) that homosexuality as a psychosexual condition became widely accepted among Western clinicians.

Also, once homosexuality was understood to be an orientation and not just a criminal behavior, it could be medicalized as a psychiatric "condition."* For almost a hundred years to follow, psychiatrists saw gays and lesbians as quite obviously mentally ill. And just as one would treat the pathological symptoms of patients suffering from any mental illness, most clinicians believed that homosexuals should be treated for their unfortunate disorder. I'll come back to "conversion therapy" in later chapters, but needless to say, such treatments, in all their shameful forms, certainly didn't involve encouraging gays and lesbians to be themselves.

The die had also been cast for the disparaging term "pervert" and its enduring association with homosexuality. Not so long ago, some neo-Freudian scholars were still interpreting anal intercourse among gay men as an unconscious desire in the recipient to nip off the other's penis with his tightened sphincter. "In this way, which is so characteristic of the pervert," mused the influential psychiatrist Mervin Glasser in 1986, "he [is] trying to establish his father as an internal object with whom to identify, as an inner ally and bulwark against his powerful mother." That may sound as scientific to us today as astrology or etchings on a tarot card, but considering that Glasser wrote this thirteen years *after* the American Psychiatric Association removed homosexuality from its list of mental disorders, it shows how long the religious moral connotations stuck around even in clinical circles. Glasser's bizarre analysis of "perverts" is the type of thing that gay men could expect to hear if they ever sought counseling for

---

*The new medicalization of homosexuality did offer some gays and lesbians a certain degree of legal protection against overzealous prosecutors who sought to jail anyone caught in a same-sex erotic tryst, especially since consensual adult homosexual acts would remain against the law for some time to come. In many places, a psychiatrist testifying on behalf of the defendant that the latter suffered from the mental illness of "inversion" could mitigate the legal punishment for such unlawful homoerotic dalliances.

their inevitable woes from living in a world that couldn't decide if they were sick or immoral, so simply saw them as both.

Today the word "pervert" just sounds silly, or at least provincial, when it's used to refer to gays and lesbians. In a growing number of societies, homosexuals are slowly, if only begrudgingly, being allowed entry into the ranks of the culturally tolerated. But plenty of other sexual minorities remain firmly entrenched in the orientation blacklist. Although, happily, we're increasingly using science to defend gays and lesbians, deep down most of us (religious or not) still appear to be suffering from the illusion of a Creator who set moral limits on the acceptable sexual orientations. Our knee-jerk perception of individuals who similarly have no choice over what arouses them sexually (pedophiles, exhibitionists, transvestites, and fetishists, to name but a few) is that they've willfully, deliberately, and arrogantly strayed from the right course. We see them as "true perverts," in other words. Whereas gays and lesbians are perceived by more and more people as "like normal heterosexuals" because they didn't choose to be the way they are, these others (somehow) did.

A subtle form of this flawed logic can even be found in the reasoning of some atheistic evolutionary biologists. When weighing in on the marriage equality debate or on other gay rights issues, many scholars like to mention the simple fact that homosexual acts are common in other species, too. This is to say, "Oh, relax, everyone, gays and lesbians are fine because, look, they really aren't *that* weird in the grand scheme of things." There's good emotional currency in animal comparisons, and I like this tack very much for its rhetorical effects. Yet it's fundamentally wrong, because it simultaneously invokes a moral judgment against those whose sexual orientations are *not* found in other animals. Furthermore, even if we were indeed the lone queer species in an infinite universe of potentially habitable planets, it's unclear to me how that would make marriage between two gay adults in love with each other less okay.

Same-sex behaviors in other species are interesting in their own right. But are we humans really that lost in the ethical wilderness that we're actually seeking guidance from monkeys, crawfish, and penguins about the acceptable use of our genitals? We engage in the same questionable reasoning when citing other nonmonogamous species to support our views on polyamorous (or "open") relationships (this was in fact a message central to the popular book *Sex at Dawn* by Christopher Ryan and Cacilda Jethá).

Even though we may be operating with the most humane intentions, when we're thinking about sex and morality, it's all too easy to fall prey to a philosophical error called the naturalistic fallacy. In effect, the naturalistic fallacy assumes that that which is natural is therefore okay, good, or socially acceptable and that which is unnatural is, in turn, bad and unacceptable. Those who use examples of same-sex liaisons in other species to justify the social acceptance of gays and lesbians are every bit as guilty of the naturalistic fallacy as the religious conservatives who perceive some sort of "obviousness" in these behaviors being morally wrong due to their "unnaturalness." After all, sex acts with reproductively immature juveniles and forced copulation are also commonly found in nature—in fact, much more so than homosexuality. Yet these more impolite details about the sex lives of other species haven't led to many moral arguments (or at least persuasive ones) for human adults having sex with children or men raping women. And they shouldn't. But we need to be careful here, because in cherry-picking features of the natural world to defend one social category—in this case, gays and lesbians—we risk shaking the whole tree.

This problem of the naturalistic fallacy is even more apparent when the "acceptable" form of human sexual deviance occurs alongside "unacceptable" forms in the very same species. Bonobo chimpanzees (*Pan paniscus*) are one of the most frequently

mentioned examples of another primate exhibiting a natural tendency to engage in homosexual acts and to have multiple sexual
partners. Given our close genetic relationship with bonobos (we
share around 98.6 percent of our DNA with them), they're often
used to showcase the "naturalness" of human homosexuality and
the "unnaturalness" of human monogamy. These apes are most
notorious for their passionate displays of "GG-rubbing" between
females (genito-genital friction involving two female bonobos
ecstatically rubbing their clitorises together), but mutual masturbation between males is also common. Such homoerotic encounters are believed to be central to the bonobo's relatively
nonaggressive nature, in that sex is used as a sort of peace offering for dialing down rising social tensions in the group before
things turn violent. As the primatologist Frans de Waal notes,
what's especially interesting is that bonobos' "sociosexual behaviors" would get humans arrested. (This is clear from his article
titled "Sociosexual Behavior Used for Tension Regulation in All
Age and Sex Combinations Among Bonobos," which was published in the journal *Pedophilia*.) Not only are "consenting adult
bonobos" being "naturally" gay and promiscuous with each
other, but you'll also find, for example, adult male bonobos
"naturally" fondling immature males alongside adult female
bonobos "naturally" mouthing the genitals of juvenile females.

When religious or social conservatives commit the naturalistic fallacy, by contrast, the issue of procreation usually takes
center stage and the philosophical error is embarrassingly salient. The trouble with confusing morality with reproduction
applies equally to any sex act that can't produce offspring, but
it's most commonly seen, unsurprisingly, in arguments against
homosexuality. Many social conservatives enjoy pointing out to
those of us who just can't seem to grasp these more challenging aspects of biology that whereas having intercourse with
the opposite sex can produce offspring, having intercourse
with the same sex cannot. This is usually extrapolated to mean

that gay sex is *obviously* unnatural, and so it's *obviously* just plain wrong (translation: arrogantly ignores God's intentional design).

Given that nature is mechanistic and amoral, and not the product of intelligent forethought, this entire position is a nonstarter. To draw from the assembly-line principles of reproductive biology any moral directive or prescription for what human beings *should* and *should not* do with their genitalia is to assume that a Creator intentionally engineered our reproductive anatomies. That pre-Darwinian view—that a preconceiving mind is needed to account for the evolution of our body parts and that this divinely penned blueprint should in turn dictate our social behaviors—lies at the malformed heart of the religious argument against homosexuality (one that's usually expressed in the form of some awful rendition of the "Adam and Steve" fundamentalist refrain).

We've become so focused as a society on the question of whether a given sexual behavior is evolutionarily "natural" or "unnatural" that we've lost sight of the more important question: Is it harmful? In many ways, it's an even more challenging question, because although naturalness can be assessed by relatively straightforward queries about statistical averages—for example, "How frequently does it appear in other species?" and "In what percentage of the human population does it occur?"—the experience of harm is largely subjective. As such, it defies such direct analyses and requires definitions that resonate with people in vastly different ways. When it comes to sexual harm in particular, what's harmful to one person not only is completely harmless to another but may even, believe it or not, be helpful or positive. If the supermodel Kate Upton were to walk into my office right now and tie me to my chair before doing a slow striptease and depositing her vagina in my face, I think I'd require therapy for years. But if this identical event were to happen to my heterosexual brother or to one of my lesbian friends, I suspect their brains would process such a "tragic" experience very differently. (And that of my not-very-amused sister-in-law would see my brother's encounter with said vagina differently still.)

It's not just overt sex acts that people experience differently in terms of harm but also sexual desires. For the religiously devout, this entire conversation is a lost cause.* But once one abandons the notion that one can "commit" a sin by thinking a thought, it becomes quite clear that sexual desires—no matter how deviant—are intrinsically harmless to the subject of a person's lust, at least in the physical sense. Mental states are "mere breath on the air," as the philosopher Jean-Paul Sartre once wrote. Sexual desires can, of course, be thought bubbles with thorns and wreak havoc on a person's own well-being (especially when they occur in the heads of those convinced such thoughts come from the Devil and yet they just can't stop having them). Still, it's only when this "mere breath on the air" is manifested in behavior that harm to another person may or may not occur.

Treating an individual as a pervert *in essence*, and hence with a purposefully immoral mind, because his or her brain conjures up atypical erotic ideas or responds sexually to stimuli that others have deemed inappropriate objects of desire, is medieval in both its stupidity and its cruelty. It's also entirely counterproductive. Research on the "white bear effect" by the social psychologist Daniel Wegner has shown, for instance, that forcing a person to suppress specific thoughts leads to those very thoughts invading the subject's consciousness even more than they otherwise would. (Whatever you do, *don't*—I repeat, *do not*—think about a white bear during the next thirty seconds.)

*It's a pity to me that a person's critical thinking would be ceded so completely to the Bible, but religious individuals point to Matthew 5:28: "But I tell you that anyone who looks at a woman lustfully has already committed adultery with her in his heart."

Our moral evaluations should fall upon harmful sexual actions with the heaviest of thuds, but not upon a pituitary excretion that happens to morph into an ethereal image in the private movie theater of someone's mind.* It's easier to agree with that statement in principle than it is to put it into practice, however. Simply having knowledge of another person's deviant desires can interfere with our ability to be so logical. If somehow you became aware that your friendly middle-aged neighbor, the one with the Honda Accord, the white picket fence, and the adoring golden retriever, derives intense sexual gratification by masturbating to violent rape fantasies, and that the mere sight of a horse's penis in one of her many glossy copies of *Equus* can bring the cheery, rosy-cheeked woman who works at the bakery up the road to a mind-numbing climax, would this alter your perception of them as respectable members of your community? Neither one, to your knowledge, has ever harmed a living soul due to his or her deviant desires (nor is there any reason to think they'd ever act on them, since they seem to have successfully refrained from doing so up to this point), yet it's still hard to keep from pulling out our moral yardstick and applying it to their "mere breath on the air."

In the real world, one reason we mistakenly conflate sexual desires and sexual behaviors in our moral evaluations of others is that the actual distance between the two can be measured by the neuron. That is to say, while it's true that sexual desires don't always turn into overt actions, it's also the case that behaviors are strongly motivated by what's on a person's mind. Using our knowledge of a person's sexual desires to morally judge him or her constitutes a philosophical error, but from an evolutionary perspective, it's the sort of bad philosophy that often leads to adaptive social decision making. Even if his history *is* unimpeachable, declining

---

*For a rare glimpse into the "average" person's private sexual fantasies, see the psychologist Brett Kahr's book *Who's Been Sleeping in Your Head? The Secret World of Sexual Fantasies* (New York: Basic Books, 2008).

an invitation from that neighbor of yours with the violent rape fantasies to join him for a romantic dinner at his place tomorrow night—just the two of you—seems prudent if you're an attractive female. And assuming I had one, I'm afraid that even I'd be slightly reluctant to let the horse-loving bakery worker near my impressive stallion.

This better-safe-than-sorry bias is morally rational (it's designed to protect ourselves or those we care about from harm), but it's not morally logical (it's a form of prejudice against someone on the basis of his or her sexual nature, not a decision based on anything the person has actually done). When it comes to high-risk social decisions such as who's going to babysit the children or whether or not to go out on a date with someone you barely know, negative stereotypes can have adaptive utility. But just because something is adaptive doesn't mean it's ethically defensible. (Assuming so would be to commit the naturalistic fallacy.) After all, it works the same way for *any* minority social category. Whether it's the color of one's skin, how much a person weighs, or the particular accent a person speaks with, negative stereotypes are shortcuts in our social reasoning that crop up mostly under time pressures and as the stakes rise. In the case of those with deviant sexual desires and our attempts to avoid harm, caution may be warranted, but we also shouldn't follow our emotionally driven intuitions so blindly that we're wholly unaware of our failure in moral logic. Overgeneralizing the frightening attributes of the most negative examples of a hidden social category (such as a religious denomination or a sexual orientation) to *everyone* in that category leads to disturbing effects in which we become overly paranoid about the unseen monsters in our midst. In sociological terms, we submit to "moral panic." Since unsavory carnal desires can't be easily detected, everyone is potentially now one of "those people."

If a measure came up on the voting ballot today for the preemptive extermination of pedophiles, I've no doubt that it would pass by a landslide. Most people, I suspect, would reason that since pedophiles are intrinsically evil, eradicating them this way is in the best interest of society. There was once a similar approach to dealing with sexual deviancy in seventeenth-century New England. You've heard of the witch hunts in Salem, but I'm guessing you're not as familiar with the pig-man hunts of New Haven. The most troubling sex fiends of those days weren't pedophiles (the age of consent in the colonies was ten, if that tells you anything) but men secretly in league with the Devil to impregnate barnyard animals. The fear was that the resulting malevolent offspring (called "prodigies"—my, how the meaning of that word has changed over time) would silently infiltrate the fledgling America and muck it all up with evil for the God-fearing folk. The settlers had gotten this strange idea from the teachings of the violently prudish medieval scholar Thomas Aquinas, who coined the term "prodigy" to refer to any hybrid creature sprung from the loins of another species but borne of human seed. According to him, prodigies could also be conceived through sex with atheists (a.k.a. perverts), but it seems there were far fewer of those milling about the colonies than solicitous swine.

It's unclear if any of the early Americans I'm about to describe were what today's sexologists would call "zoophiles," individuals who are more attracted to nonhuman animals than to human ones. They may have merely used members of other species as surrogates for human partners in obtaining sexual gratification (as half of all "farm-bred" adolescent males have done, according to Alfred Kinsey in 1948), or they could have been falsely accused of such acts altogether. Yet some modern scientists believe that zoophilia is a genuine sexual orientation represented by as much as a full 1 percent of the human population. Just as it's impossible for nonzoophiles to become aroused by the steaming, mottled member of a Clydesdale or by a German shepherd rolling over for a tummy rub, "true zoophiles" can't get (easily) turned on by human beings. One such man—a physician from suburbia, incidentally—could only consummate his marriage to a woman by closing his eyes and imagining his new bride as a horse. Strangely enough, the marriage didn't last.

Centuries ago in the newfound colony of Plymouth, zoophilia was obviously not a known sexual orientation (again, the psychosexual construct of an "orientation" wouldn't appear until the late nineteenth century). But the hysteria over Satan's prodigal litters reached dramatic heights with the 1642 trial of a sixteen-year-old boy named Thomas Granger. This randy adolescent had been indicted for taking indecent liberties with what seems an entire stable full of animals, including "a mare, a cow, two goats, five sheep, two calves and a turkey." I realize the turkey part is a bit distracting (and how one goes about having sex with a large clawed bird is better left unexamined), but even more remarkable is the legal diligence and sobriety with which this case was prosecuted.

There was little question in these righteous minds that the boy should be dispatched to the flames for his egregious violations of natural law, but there was confusion on the bench over which sheep, exactly, he'd been defiling, and therefore which of them should be killed and which of them spared. This was crucial to sort out, not only because livestock was a valuable commodity in the beleaguered settlement, but also because if they executed the *wrong* sheep, they risked the unthinkable happening: a monstrously bleating, hoofed prodigy might drop undetected onto Plymouth. So, naturally, a lineup of busily masticating victims was staged for Granger. With a trembling finger, the boy pointed out those five amber-eyed ruminants that had been targets of his secret woolly lust. Court records indicate that the animals were then "killed before his face, according to the law, Leviticus xx. 15; and then he himself was executed."

Suspected "buggers" of the past—Old English slang for he

who has sex with pigs, donkeys, dogs, and all and sundry crit-ters—could expect a battle of wits with moral arbiters and over-zealous prosecutors. In New Haven in 1642, not far from what would later be the Yale campus and just a few years after the Granger affair, a man named George Spencer, a servant notori-ous for having "a prophane, lying, scoffing and lewd speritt," was executed for making love to his master's pig. He swore that he didn't do it, but unfortunately for Spencer the sow happened to give birth to a deformed fetus ("a prodigious monster") that re-sembled George a bit too closely for most people's comfort. The pig fetus had "butt one eye for use, the other hath (as it is called) a pearle in itt, is whitish and deformed." This embryological mis-hap was George's death sentence. His own ocular deformity bore an uncanny resemblance to that of the stillborn pig, and in the emotional climate of the town's moral panic over grunting prodi-gies, this was the critical piece of evidence used to convict him.

Another town resident with the rather ironic name of Thomas Hogg also found himself at the center of an intense buggery in-vestigation when a neighborhood sow bore a deformed fetus with "a faire & white skinne & head, as Thomas Hogg is." (I feel compelled to pause for a moment to pity the women of old New Haven, too, since so many aborted pig fetuses were apparently reminiscent of the town's eligible bachelors.) The allegations made against Thomas Hogg by the townsfolk were so serious that the governor and the deputy governor personally frog-marched him out to the barnyard toward the sow in question and or-dered him to "scratt" (fondle) the animal before their eyes. This was done to gauge just how intimately familiar they might be. "Immedyatly there appeared a working of lust in the sow," the court records recount, "insomuch that she powred out seede before them." When Hogg reluctantly titillated the teats of a dif-ferent sow, by contrast, she didn't return his affections. At least, she didn't release her bladder when he touched her teats, which is probably what that earlier pig had done. Given his accusers' unimpressive knowledge of biology, she seems simply to have peed at the worst of times. So Hogg, like Granger and Spencer before him, was executed.

These zero-tolerance laws against bestiality had been imported from Christian Europe, the stomping ground of zealots like Aquinas. But an interesting development emerged on that side of the Atlantic in the eighteenth century, highlighted by the case of a ragged French peasant named Jacques Ferron, who was tried for having sex with a female donkey. As described by Edward Payson Evans in his 1906 cult classic of legal scholarship, *The Criminal Prosecution and Capital Punishment of Animals*, Ferron would clearly be killed, since he was "taken in the act of coition" with the animal. He'd soon be shoved along in shackles to the public square, where a smoldering stake was waiting to consume him in flames as he pleaded for mercy against a wall of scornful faces. What makes Ferron's case different from the bestiality trials up to then is that the locals chose not to slay the jenny along with him. In fact, the donkey was so beloved by the community that she was instead given her own separate trial, with witnesses to testify that not once had they ever seen her exhibit even the slightest sign of promiscuity. Before the proceedings, a certificate was even drawn up affirming the donkey's virtuous reputation. This impassioned plea was signed by the parish priest and was enough to persuade the court officials to acquit the animal on the grounds that she'd quite clearly been raped.

This French donkey-rape case may sound somewhat absurd to us today. But it was a small moment in history in which people stopped and questioned the punishment demanded by the Bible and instead chose their own more rational course of action, showing how even a society steeped in religion can move away from the irrelevant question of *naturalness* and onto the more meaningful and moral one of *harmfulness* in its consideration of sexual deviance. Personally, I don't think Ferron got a fair shake, since harm to the animal hadn't really been established. Most of us (me included) don't especially enjoy the thought of a man screwing a donkey, let alone such an apparently virtuous one, but anyone who has ever seen the erect penis of an adult jackass, approximately the size of a small moped, would have to acknowledge that it's unlikely Ferron's member caused physical injury to the donkey. And unless she stopped eating, was ostracized by the group, or felt ashamed by the judgmental glares of the other donkeys, psychological damage was also unlikely. Still, since God clearly prescribes death to any creature, willing or unwilling, tainted by human semen, the sparing of this she-ass meant her *perceived* harm (by rape) was important enough to these people to ignore God's unreasonable and cruel orders to kill her. They thought for themselves in this sexual ethical dilemma, in other words. And that's real moral progress. (True, they still chose to burn the human being alive. But—baby steps.)

Since we now know that many people of European descent possess Neanderthal DNA, it's tempting to speculate about how many of those stoking the fires of yore were flesh-and-blood prodigies themselves. My Heinz 57 genome has never been sequenced, but there's a good chance I'm a hybrid in this sense, too. (*See*, my adolescent caveman crush was perfectly "natural.") But in any event, once some basic biological knowledge put an end to the paranoia over Aquinas's evil prodigies not long after the Ferron affair, Christians abandoned the practice of immolating those accused of interspecies sex.

It's foolish, however, to assume that religious morality isn't still woven into modern bestiality laws (even the term "bestiality" is religious, first appearing in the King James Version of the Bible in 1611 to falsely cleave apart human beings from all other animals). We're a peculiar species, but humans are animals too, of course. Bestiality is expressly illegal in most countries today, and in those places where it's not an officially codified crime, people who have sex with animals are still occasionally prosecuted under animal cruelty laws. As a *platonic* animal lover, I'm in favor of protective laws. The sad reality is that there are indeed hideous sexual deeds done to animals by a few demented people. Yet there are also cases of human-animal sex that don't involve any obvious harm to the animal and may even involve mutual pleasure. Which is worse, for instance, a stud manager forcibly collecting the semen of a prized racehorse by "electro-ejaculating" the animal for commercial gain (which involves inserting an electrified rod into the animal's rectum and delivering a high-voltage shock to its prostate) or a zoophile gently masturbating his companion horse with the sole intent of bringing it satisfaction? That the first is perfectly legal and the second illegal shows that bestiality laws are more concerned with a person's sexually deviant desires than they are with the animal's actual harm. When the question of harm is an afterthought in any sex law, we need to rethink both its fairness and how it's handled by the courts. There *is* the problem of an animal's inability to give verbal consent. But note that many zoophiles prefer to be the passive *recipients* of the animal's actions upon—or more often inside—them. That's still completely illegal, even in cases of volitional thrusting (think humping dog to human leg), which seems to imply the animal is more or less on the same page with the zoophile. Funny enough, that equally thorny problem of how to gain an animal's verbal consent before it's killed for one's personal dining pleasure doesn't inspire nearly the same degree of outrage. Not that either is great, but if I were a bovine, I'd rather get "humanely" penetrated by the penile equivalent of a stiff strand of hay than be "humanely" slaughtered by seventeen-inch steel blades. (I think I might have to draw a firm line with all this at juvenile goats, though. They're just kids, for God's sake.)

❧

Most modern sex crime laws are based on the better-safe-than-sorry principle. And that's quite sensible. We especially want to protect the most vulnerable members of our society—children, animals, the elderly, the disabled—so we err on the side of extreme caution, even if this means occasionally getting it wrong about the actual harm that they face. After all, what's the life of one well-meaning, gentle sex deviant when it comes to protecting those we care about? Lumping that unfortunate sap in with the nastier variety is a chance you're probably willing to take. Once you accept that we're all sex deviants in our own ways, however, the life of that expendable pervert is more than just collateral damage. That unfortunate sap could, in fact, be *you* someday, or the very child you're trying to protect. Given the serious consequences of our "true natures" being known, is it any wonder that your first thought (and I'm paraphrasing here) was "Speak for yourself" when I initially called you a pervert?

It's far easier to assume that all sex deviants, including even some of those who've committed crimes, are immoral than it is to show, case by case, how they've caused measurable harm. "In all the criminal law," Alfred Kinsey once pointed out, "there is practically no other behavior which is forbidden on the ground that nature may be offended, and that nature must be protected from such offense. This is the unique aspect of our sex codes." Once the Bible and the legal system aren't there to tell us how to think about sex (neither of which, you may have gathered, I'll be using as a source of moral authority for our consideration in the chapters ahead), establishing harm can require strenuous mental effort. In fact, you might think you're pretty good at knowing sexual right from wrong. But even our most fervent intuitions aren't always as logical as we'd like to believe.

Back in 2001, the psychologist Jonathan Haidt coined the term "moral dumbfounding" to refer to the phenomenon in which we struggle to elaborate on the precise reasons why we believe certain acts are immoral. Emotionally fueled tautologies (or expressions of redundancy that fail to offer any actual clarification, such as "It's wrong because it's just nasty," "You shouldn't do it, because it's creepy," "It's immoral because it's plain evil," and of course "It's not right, because God says so") only echo intense social disapproval for certain crimes that shouldn't be crimes at all when we prioritize the question of harmfulness. Consider a vignette from a study in this area:

> A man belongs to a necrophilia club that has devised a way to satisfy the desire to have sex with dead people. Each member donates his or her body to the club after death so that the other members can have sex with the corpse. The man has sex with a dead woman who gave her body to the club.

When asked whether it was wrong for this man to do what he did and, more important, to articulate and to justify their belief if they said it was, most participants in this study defaulted to a *presumption of harm* in their moral reasoning. Even when they were told explicitly that the woman didn't have any family members who might get upset if they found out what happened to her corpse, that the club isn't interested in recruiting or harming living people, that neither the man nor any of the other club members suffer any regrets or anguish about their sexuality, that the group's activities are kept private and consensual, that the man used protection to prevent disease, and, per her instructions, that the club cremated the woman's body after the man was done having sex with it, people *still* insisted that somehow or another, someone, somewhere, must be getting harmed.

For social conservatives, the damage might even be seen as inflicted on symbolic bodies—"America," for example, "the Church," "society," or "the sanctity of marriage." Saying that a behavior is "harmful to America" or that it's "destructive to society" is a bit like giving corporations the legal status of personhood. That is, it only makes sense to those with an agenda. The scientific definition of a "person" as a carbon-based life-form resembles nothing of the circuitous legal definition that enables a profit-driven corporation to claim that same status. Likewise, pain and distress can occur only at the level of a subjectively experiencing organism (human or animal) in possession of pain receptors and a nervous system able to register emotional trauma, *not* at the level of an abstract entity without a brain. The problem of sexual harm concerns living, breathing creatures, not political parties, nations, or worldviews.

The case of the responsible necrophile is just one example of deviant sex in which researchers have uncovered presumption-of-harm reasoning. When offered similarly clear information about eliminating *all possible* forms of measurable harm, participants trying to justify their feelings of the wrongfulness of sex with animals, teenagers, and family members (incest) likewise default to a presumption of harm. Such scenarios aren't confined to artificial lab studies, either. They happen in the real world as well. Indeed, that's the whole point: that not every "obvious" case is in fact so obvious.

Take the brothers Elijah and Milo Peters, for example, a pair of twentysomething identical male twins from Prague who appear together in gay porn films featuring full anal penetration—*with each other*. The Peters twins not only have been having sex together since they were fifteen but also consider themselves romantic partners, just like other young couples with genes that don't match so perfectly. Outside the porn studio, they claim to be monogamous. "My brother is my boyfriend, and I am his boyfriend," says one of the other. "He is my lifeblood, and he is my only love." With the procreation factor removed (and therefore the possibility of genetic harm to any resulting offspring able to be completely ruled out), along with the Peters twins giving mutually enthusiastic consent to sex, their surprising absence of shame about it, and their clear happiness with each other, their steamy incestuous pairing isn't so obviously "wrong."

One reason it's so difficult for us to exercise our mental faculties in a proper way when it comes to the subject of deviant sex, instead being ruled by emotional reactions that fail to give accurate weight to the question of harm, is what we might call "the disgust factor." Feelings of disgust have a way of undermining our social intelligence and indeed of compromising our very humanity. In fact, as we're about to see, if there's anything that researchers have learned about moral reasoning and sex in the past decade, it's that disgust is the visceral engine of hate. The good news is that once you understand how the whole thing works, you can kill that engine. Our best hope of addressing this deep-seated problem of sexual disgust is to do some reverse engineering on its adaptive functions. Because let's face it, when you're not in the mood or you're not attracted to the person whose sex life it is that you're contemplating, sex can be gross. And deviant sex, almost by definition, is bound to gross out more people than normal sex. But disgust doesn't justify the ravages of inequity and oppression on the lives of sexual deviants themselves.

TWO

DAMN DIRTY APES

The butting of his haunches seemed ridiculous to her, and the sort of anxiety of his penis to come to its little evacuating crisis seemed farcical. Yes, this was love, this ridiculous bouncing of the buttocks, and the wilting of the poor insignificant, moist little penis.

—D. H. Lawrence, *Lady Chatterley's Lover* (1928)

The entire ordeal is something of a blur to me now, but the one thing that I remember clearly about my first experience with another man (a real *Homo sapiens* this time) is that he was far more interested in fellating my toes than he was in doing anything with some other body part of mine. Well, different strokes for different folks, you'll say. Really, that's quite kind and understanding of you. But if you ever have the misfortune of actually seeing my feet, which are vaguely reminiscent in both color and shape (I hesitate to say smell, but if truth be told, sometimes that too) of the sparsely haired underbelly of a dead possum, you'd realize just how extraordinary this man's bedroom behaviors really were. That a person could become so sexually excited—in the full curtain-drawn light of day, no less—by something that I perceived to be so *disgusting* mystified me.

To this day, I avoid making direct eye contact with my feet when taking a shower, so it's still hard for me to completely understand his actions. I do, however, have a better sense of the mechanics behind this man's lustful psychology. First of all, it's clear he was a podophile. The words look and sound very similar, but note that's an *o* and not an *e* as in "pedophile." (I was young but not that young, after all.) Podophilia, or "foot fetishism," is by far the most common manifestation of what sexologists refer to as a sexual "partialism," which is an erotic preoccupation with a nonreproductive body part. Feet, belly buttons, teeth, noses, eyeballs, earlobes, pinkie toes, calves, nipples—there are partialists for any type of localized real estate you can imagine, and their desire for this part exceeds (and sometimes even excludes) their interest in the genitalia. In any event, my awkward first experience with a disgust-challenged podophile who was willing to be intimate even with *my* feet encouraged me to read up on the curious history of foot fetishism.

It was none other than Havelock Ellis who first unraveled the mind-set of the podophile.* Unlike the subject of *Sexual Inversion*, Ellis's sharp-eyed analytical focus on foot fetishists zeroed in on the heterosexuals among them. "In a small but not inconsiderable minority of persons," he wrote in 1927, "the foot or boot becomes the most attractive part of a woman, and in some morbid cases the woman herself is regarded as a comparatively unimportant appendage."† I know how she feels. Ever since Ellis dug

---

*Or what Ellis called "retifism," after the eighteenth-century French novelist Rétif de la Bretonne, who wrote that as a boy he would "tremble with pleasure" and blushingly lower his eyes before a pretty girl's boots "as if in the presence of the girls themselves."
†I'll go into more detail about the developmental origins of fetishes in a later chapter, but in a study with gay male podophiles who were members of the "Foot Fraternity," the sociologist Martin Weinberg found that most of the men could trace their passion for feet back to a specific childhood event. Quite often, these were innocent experiences of playing with their parents' lower extremities. "Sleeping upside down with my parents," reflected one man, "and finding my dad's feet in my face." "I used to tickle my dad's feet," recalled another. "I enjoyed his laughter very much . . . he would feign enjoyment as part of the game." Another reminisces: "At about five or six years old, remov-

his heels into the matter, case studies of foot fetishism have continued to find an attentive audience. Homosexual, heterosexual, and even bisexual podophiles have all made sporadic appearances in psychological write-ups. But as far as I know, there has only been one attempt in all of podophilic history to explain foot fetishism using evolutionary theory. And believe it or not, it's not an altogether ridiculous Darwinian hypothesis, either.

The psychologist James Giannini put the idea forward in 1998. Giannini had discovered a revealing sociosexual trend concerning podophilia. Throughout the course of human history, the cultural eroticization of the female foot predictably peaked whenever there was an outbreak of venereal disease, and then just as predictably it leveled out again as the epidemics ran their courses. Foot love blossomed during the gonorrhea epidemic in the thirteenth century, for example, syphilis in the sixteenth and nineteenth centuries, and even AIDS in the late twentieth century. (As if the oppressive Inquisition weren't bad enough, Spain was also suffering from a large syphilitic population just as the heresy trials were heating up. With all that was going on, it seems like an odd time for Spanish painters to begin specializing in portraits of women's feet, but this is precisely when that artistic oeuvre really took off. New shoe styles showing a teasing bit of "toe cleavage" were also all the rage.) Even if you're straight and

---

ing my father's shoes and massaging his hot feet . . . the soft, warm feet and the pleasure he seemed to experience—usually going to sleep—and I could kiss and lick his feet." See Martin S. Weinberg, Colin J. Williams, and Cassandra Calhan, "'If the Shoe Fits . . .': Exploring Male Homosexual Foot Fetishism," *Journal of Sex Research* 32, no. 1 (1995): 17–27. These men just happened to be gay, but it works the same way for straight podophiles. In a detailed psychoanalysis of a "child foot fetishist," a team of neo-Freudian sleuths tried to unravel the case of a sixteen-year-old boy who'd been enamored with his mom's feet since he was a toddler. She originally thought it was cute, but by the time he was six, his fascination with his mom's extremities had become sexualized. "While licking the feet," write the psychiatrists, "he regularly had an erection and played with his penis." See Jules R. Bemporad, H. Donald Dunton, and Frieda H. Spady, "The Treatment of a Child Foot Fetishist," *American Journal of Psychotherapy* 30, no. 2 (1976): 303–16.

into a lady's lower extremities, you can't very well impregnate her foot to spread your genes. Giannini's claim was simply that if one's arousal were *primarily* but not *exclusively* confined to nonreproductive parts, then less frequent contacts (or maybe less exuberant ones) with infectious genitalia could meaningfully reduce the risk of infertility or even death. If such outbreaks were common enough in our deep past, suggested Giannini, then people who were able to become sexual partialists would have an advantage over those concentrating all their attention on the body's more dangerous hot spots.

There's still that puzzling question of how the podophile could suckle toes from hooves as hideous as mine. I do try to keep them clean, but they are feet, after all, and one can't always know exactly what's going on down there with the fungus scene. In fact, never mind feet, it's astonishing that we're so willing during sex acts to put *any* body part in our mouths that doesn't belong to us. Penises don't always come out smelling of roses either, and consider the flourishing bacterial substrate that is the human vagina. This region can play host to more than four hundred different species of organisms, and healthy female anatomies contain numerous acids that combat yeast and pathogens and give vulvae their odoriferous punch.* Not only that, but in both sexes, there are distinct glandular secretions that you'd rather

*One group of enterprising scholars isolated the "vaginal vapors" of ten different women to see if all vaginas smelled alike. There were some strict rules that the open-minded women who volunteered for the study had to follow: no douching for a full week before the scent sampling, no intercourse for forty-eight hours prior, and absolutely no garlic or heavy seasoning in their foods, since these substances leak into genital fluids. The sniffers' ability to differentiate the vaginal odors by female donor led the authors to conclude that, indeed, depending on the particular mélange of inhabitants occupying an unwashed groin, every woman on the planet has her own signature smell. See Louis Keith, Paul Stromberg, B. K. Krotoszynski, Joan Shah, and Andrew Dravnieks, "The Odors of the Human Vagina," *Archives of Gynecology and Obstetrics* 220, no. 1 (1975): 1–10.

not know about gathering unseen around the anus, face, groin, scalp, and umbilicus. There's also, of course, prodigiously generated sweat, tears, urine, dental plaque, sebum, earwax, smegma, and that most formidable foe to our sexual arousal that is feces. More specific culinary hurdles depend on the sex of your partner. If employing your mouth on a man's body, for instance, your palate can anticipate being greeted unexpectedly by pre-ejaculate or semen. Women's equally aqueous bodies, by contrast, are often plentiful reservoirs of vaginal fluids, breast milk, and menstrual blood. Considering what walking factories of ick we human beings are, it's amazing that we've managed not only to *survive* as a sexual species by wanting to copulate with each other but to do it often enough over our 150,000-year eyeblink of an existence that we're now straining the planet's natural resources beyond all capacity.

The secret to our "success" lies in how our lustful mammalian minds evolved to handle each other's sometimes-repellent bodies. It's quite an exquisite operating system, too. Lust and disgust are antagonistic forces in an emotional balancing act that serves to push us toward orgasm (through lust) or to turn us away from it (through disgust). It's a dynamic relationship with ancient origins. For example, DNA sequencing reveals that the "murid rodent ancestor" (a term that signifies the last common ancestor of human beings, mice, and rats—so a slightly different take on Steinbeck's *Of Mice and Men*) last scurried upon this earth around eighty-seven million years ago. Yet the subjective experience of disgust is every bit as much the modern rat's carnal kryptonite as it is ours. If you take a healthy, virile adult male rat and allow him to have unbridled sex with a female in heat and then immediately inject him with a nausea-inducing drug such as lithium chloride, he'll acquire a total aversion to sex. Nothing else has this effect. Even if you shock him while he's doing it, or inflict any other type of cruel punishment, this won't diminish his sexual appetite—only disgust does that. It's just his mating behaviors that are affected, too; the lithium chloride has no effect on his social behavior in general. He'll be as affable with his rat friends as ever, in other words, but he won't be doing that horrible pelvic-thrusting stuff for a while, since that's what made him so miserable the last time around.

While we're on the general subject of our animal ancestry, one of the more creative accounts of the relationship between sex and disgust in human beings comes from the field of "terror management theory," which postulates that any disgust reactions we have to sex actually stem from the fear of our own mortality. Sex is so corporeal, or bodily, the argument goes, that it's a uniquely powerful reminder of our animal nature. And just like other animals we've got one-way tickets to Decomposition Central, which is a very scary place to us. It's so scary, terror-management theorists claim, that if our brains were to dwell on this reality for too long, we'd become so paralyzed with fear that we'd no longer be able to function adaptively. Human beings coped with their awareness of death, these scholars believe, by inventing various cultural expressions of immortality to quell their existential fright. (All of this is presumably happening subconsciously, mind you.) And by the looks of it, the idea of sex presents some big challenges to our species in this uncomfortably mortal regard.

In one study, for instance, having people contemplate their own deaths caused them to favor a definition of sex in its more lofty, abstract forms (such as "making love" rather than "copulating"). Concepts like love and romance are said to be "symbolically immortal," helping to return the individual to a more manageable state of death anxiety. Presumably, this is why we're so enamored of slogans such as "A diamond is forever" and why poetic lines like Emily Dickinson's "Unable are the Loved to die / For Love is Immortality" strike a true and universal chord. Compare those maudlin sentiments with Shakespeare's lurid metaphor for sex in *Othello* as "making the beast with two backs" or his describing a couple as "prime as goats" and "hot as monkeys." You'd find Shakespeare's language here especially gross, the argument goes, right after being told you've got an inoperable brain tumor and have only a few months to live. Like another famous theory that's based on subconscious anxieties, it has its limits, but I think there's probably something to terror management theory. It could explain Rick Santorum believing gay marriage is a stone's throw away from interspecies marriage, anyway. That poor man—the mere act of ejaculating, reminding him that he's an animal, must be so existentially *terrifying* to him.

Yet we don't need complex psychodynamics to understand how it all really works. We lost our rodent tails and whiskers long ago, but even for us postmodern animals it's still best not to think *too* much about all those bodily secretions and worrisome odors of ours while having sex. Even the best relationships can be strained by having to explain to your loving partner why you look as if you've just gotten a nose full of rancid vinegar while performing oral sex on him. Fortunately, most of us can overcome these seemingly insurmountable sensory barriers thanks to a brain that's able to subjectively sanitize such sticky situations. One that evolved to be overly sensitive to disgust during sex wouldn't have been very adaptive after all; so prudish a creature would perish in its own purity as a genetic dead end. Rather, as Sigmund Freud wrote, "our libido thrives on obstacles" and "in its strength enjoys overriding disgust." Indeed, more recent scholars have found that our willingness (and sometimes our eagerness) to let others' body products make contact with our lips and tongues or even to slip down our throats entirely is a product of our fluctuating arousal levels. When we're horny, we're happy to dip into someone else's organic buffet. Really. There are even data on it.

In a study with straight undergraduate students in Denmark, for example, most of the males said they'd be willing to taste a woman's breast milk if they were aroused, but far fewer said they'd ever do so if they weren't turned on. Similarly, most women could see themselves ingesting semen if they were hot and bothered, but the thought of swallowing fresh seminal fluid when not in the mood was enough to send shivers up many spines. Imagining having to taste someone else's sweat, tears, and saliva, meanwhile, wasn't especially sickening to the students either way. By contrast, very few men *or* women—no matter how concupiscent they might be—ever wanted to taste menstrual blood. And wouldn't you know it, while most of the female students said they'd be willing to give it a go if they felt sufficiently lustful, only 3 percent of the males in the study warmed up to the idea of ever tasting another man's smegma.

This overall pattern of findings makes some sense in evolutionary terms. When the system works smoothly, sexual arousal can serve to anesthetize the otherwise adaptive disgust response long enough for people to get on with the Darwinian business of reproduction. (Well, heterosexual people, anyway. It just allows us homosexuals to have decent sex.) But questionnaire studies aren't the greatest when it comes to tapping into behaviors fueled by strong emotions. Even if a straight man *were* willing to taste another man's smegma when drunk with desire, he might not know it himself, or he might not share that fact on a survey. More creative approaches induce actual sexual arousal in participants to see how the state of lust alters their real-time perception of disgusting objects or affects their behaviors. In one study, researchers first exposed straight male subjects to porn. Only then did the real testing begin. The psychologists behind the scenes, led by Richard Stevenson, were curious to know if sexual arousal lessens disgust for *sexual cues only* or for *nasty things altogether*. We can therefore think of their experiment as pitting a "local anesthetic" hypothesis of disgust management against a "general anesthetic" one. To get at this, the scientists invaded the sensory systems of these now very aroused men with terribly yucky things, comparing their perceived grossness of gross sex cues with their reactions to cues that were just generally gross.

For the sensation of touch, for instance, the lustful men were told to dip their hands into a bucket filled with either lubricated condoms (for sexual disgust) or cold pea-and-ham soup (for general disgust). For hearing, they listened to a brief sound bite of someone either performing oral sex or vomiting. (These do sometimes go together, but let's not complicate matters.) Smell, meanwhile, was a whiff of rotting fish (for sexual disgust) or of feces (for general disgust). Seeing involved being exposed to the image of a horribly disfiguring scar on a naked woman or a pile of decomposing garbage. Fortunately for the participants, the researchers chose not to include the sense of taste in this particularly interactive study of disgust. That would have probably been "a little much," as my dad likes to say.

The findings support our local anesthetic hypothesis. Sexual arousal, or at least male sexual arousal, numbed the participants' gut-level aversion to sexual unpleasantries only, not to the whole disgusting lineup. Even when we're in the middle of doing disgusting things with each other under the sheets, in other words, we're just as sensitive to the revolting features of the noncarnal world above. Think of it this way: When you're in the throes of orgasm, the fact that the hot guy or gal you met at the club last night hasn't showered since yesterday and smells a little funky probably isn't *too* much of a hindrance to your ultimate pleasure. But if you move your passionate lovemaking to the other room, where in your mad dash to tear each other's clothes off, your nostrils suddenly detect the unmistakable smell of decomposing human flesh coming from beneath the bed, that would probably be a sensory deal breaker for your bliss. (Not to mention you'd have a different type of "happy ending" to worry about now—in the form of getting out alive.) In fact, the olfactory category in Stevenson's study, which, if you recall, compared the rotting fish smell with the feces smell, was the only one for which the lustful male raters found the stimuli to be equally pleasant (or unpleasant, depending on how you look at it). But that's actually not very surprising if we understand lust as dampening sex-related disgust

for *all* bodily effluvia, not just those emanating from one particular orifice. Necrophiles aside, the smell of a moldering corpse
tends to be sexually off-putting. Yet there are plenty of smells
from living bodies that most of us don't find too pleasant, either.
And it's not just "fishy" smells that can be an orgasmic barrier for
straight men, after all. Women, from what I understand, have
anuses too, and an *odeur de colon* can be a psychological im-
pediment to sex for both men and women, gay or straight.

By virtue of certain "anatomical affordances," however, and
the limiting nature of homosexual male intercourse (that is, anal
sex), gay men are frequently the targets of a cheap rhetorical
strategy designed to evoke a moralizing disgust response. "It's
wrong because it's gross" is, of course, a rather transparent case
of moral dumbfounding. Yet painting gay men as depraved creatures rife with infectious disease whose idea of a relaxing Sunday
afternoon involves wallowing in feces and smelly assholes can be
alarmingly effective at keeping heterosexuals from seeing them
as fellow human beings. For example, whenever the subject of
gay men crops up at the various websites, news feeds, and online
forums that cater primarily to social conservatives (they do *really* love to talk about us queer folk), post after scatological post
is a display of this antifecal-therefore-antigay mentality. "If your
personal identity revolves around your lust for other men's stink-
ing anuses," wrote one man at the *Free Republic* website in response to a news story about gay pride, "a particularly disgusting
form of depravity that spreads horrific diseases, the chest swells
with self-satisfaction." Another weighs in: "Homosexuals should
never have got their special rights like civil unions let alone marriage or Don't-Ask[-Don't-Tell] in the first place. The ignorant and
stupid on our side wanted to appease them and usually say: 'I know
a couple and they are nice' . . . friggin idiots . . . when did we base
special rights based on fecal diseased sex in the constitution?"

When the romantic relationships of gay men are repeatedly
linked with excrement this way, the disgust response makes it
much easier to sell them as immoral. And in the battle against

the "homosexual agenda," this has quickly become a core strategy in the tireless and patriotic efforts against, ahem, "evil." Really, I'm not kidding: anything even remotely related to gay men—Anderson Cooper chuckling uncontrollably during a newscast, Barney Frank bending over to tie his shoelaces, a scandalous plot twist on the sitcom *Modern Family*—will faithfully serve to induce a reverie of maledictions for their anuses.

We can also induce disgust to sway a person's political views. It's one of the oldest tricks around. "Just look at these guys!" the Nazi author of an illustrated children's book wrote for his young German readers in 1938. "The louse-infested beards! The filthy, protruding ears, those stained, fatty clothes . . . Jews have an unpleasant sweetish odor. If you have a good nose, you can smell the Jews." (The book's publisher was later executed as a war criminal for his central role in peddling anti-Semitic propaganda.) For those who don't know how the trick works, it can get them every time. A study by the psychologists Yoel Inbar, David Pizarro, and Paul Bloom illustrates how the experience of disgust can make even those who aren't normally intolerant express strikingly bigoted views. In the study, a group of straight male and female undergrads from Cornell University were randomly assigned to one of two conditions. For the sake of clarity, we'll call these the "stink" and "no-stink" conditions. All of the students sat alone in a private room and completed a survey about their attitudes on a range of social and political issues. For those in the stink condition, a research assistant had sneaked in beforehand and applied a novelty spray to a trashcan in the corner. (This smell—I got the sense from speaking to one of the authors—was vaguely akin to that of a hospital emergency-room toilet.) The no-stink participants were tested in the same room and answered the very same survey, but, luckily for their brains' olfactory bulbs, they weren't exposed to the bad odor while doing so.

The findings are revealing. Irrespective of the gender of the participants and their self-reported political orientation (they ranged from "extremely liberal" to "extremely conservative" in their worldviews), those who were randomly assigned to the stink condition expressed significantly more disapproval toward gay men than did those in the no-stink condition. It wasn't that the bad smell simply made these participants crankier in general while pondering their feelings on social issues. In fact, their opinions of other minority groups (such as blacks and the elderly) weren't any different from those of the cohort in the no-stink condition. The disgusting smell just made these very bright and perfectly polite students from an Ivy League university less sympathetic to gay men. (The data trended in this direction for their judgments about lesbians, too, but the effect wasn't as strong.)

Now, we needn't dwell on the obvious (anal sex isn't limited to gay men, not all gay men have anal sex, straight people can get HIV through intravaginal sex, and—it's really not a myth—condoms *do* exist). But just for the record, the average gay man is no less repulsed by a sexual experience involving feces than are most other human beings. I'm careful here to say "most" because there are indeed a handful of people who walk among us who could be considered truly "coprophilic," which means they have a strong erotic attraction to feces. It's not my sort of thing, and I'd imagine there's a lot more scrubbing involved than there is with your average postcoital cleanup job, but assuming one takes certain precautions, it seems harmless enough to me, really. Kind of a funny thing about coprophilia, too: the only documented cases are heterosexual men.

Feces ranks among the most rare aphrodisiacs, but we shouldn't forget that rather than seeing it as a turnoff, some people actually get turned *on* by the human body in its more natural aromatic form. Upon finishing a military campaign, Napoleon Bonaparte, who despised the saccharine fragrance of perfumes, sent that notorious memo to his wife, Joséphine. It read simply:

"Coming home soon; don't wash." The existence of human pheromones (or scent-based hormones that stimulate strong desire in the opposite sex) is still hotly debated, but whatever their role in our sexuality, it's no longer as major as it was in the days of the murid rodent ancestor. Male hamsters, for example, will hump anything dabbed with the vaginal secretions of a female in heat. By contrast, human males told to inhale a woman's bottled-up vaginal odors without knowing what it is that's under their noses generally find it off-putting. According to the psychologist Roy Levin, most heterosexual men express an "affectionate distaste" for the smell (tolerating it but not exactly enjoying it). Yet he also notes that plenty of other men, and lesbians too for that matter, have come to fetishize it. Certainly there are brazen admirers, especially when the smell is indicative of youth. Levin describes stumbling upon a product called Girl Scent on the Internet (now be nice, it was during his very important research *obviously*) with this colorful marketing description:

> Girl Scent is a new erotic scent scientifically designed to smell like a young woman's sexy vagina aroma. It was created by studying the sex pheromones and other aroma chemicals in samples of vaginal secretions gathered from dozens of healthy girls. Many men risk disease by purchasing worn panties on the Net. Girl Scent gives you a whole bottle without risk of infection!

"There does not appear to be a similar scent of male genital odor synthesized for women," observed Levin.

That may be. But perhaps it's only so because we don't yet have data on women smelling semen or men's crotches. Rest assured, however, that we do have findings on them inhaling their male partner's armpit secretions. And some women report an intense olfactory appreciation of this distinctive scent. This is particularly the case when men have different major histocompatibility complex (MHC) alleles (or genes that code for proteins involved in the recognition of biological related-ness) from the woman. When these MHC alleles differ, women judge male scents to be more pleasant (or at least less heinous) than when the alleles match. Evolutionary biologists believe that this is an adaptation—an unconscious one, needless to say—meant to steer women away from incest. Any prospective gene-tic recombination involving these different alleles helps to insulate offspring against disease and recessive mutations. If you're a straight woman of reproductive age and dare to put the theory to the test, have your dad and your current male sex part-ner (husband, fiancé, boyfriend, the cute-but-not-very-bright barista who's been giving you those courtesy espressos lately, any of them will do, really) wear the same undershirt for a full, un-bathed week. Once the two men peel these items off their ripe torsos, code whose is whose somehow and then mix them up so you can't tell by appearance alone. Dab the armpit portions of both malodorous garments generously on your philtrum (that little odor-catching trap between your nasal septum and your upper lip) and see which makes you amorous. And if neither of these mephitic fabrics gets you worked up, don't worry: note that sometimes it isn't so much a pleasure response as the lesser of two evils that reveals the general effect.

Even for those of us who are prone to gagging at basic human emissions, however, our locally anesthetized disgust usually helps us to get the sexual job done.* Still, in the heat of the mo-

---

*There are notable individual differences in disgust sensitivity. Consider some exam-ples from an assessment scale developed by the psychologist Peter de Jong. How willing would you be to "lie beneath bedclothes below which you have masturbated the day before and which show obvious smudges" or to "touch a soiled, unwashed towel that is possibly used to wipe off sperm/vaginal fluid of an unknown person after sexual inter-course (e.g., a towel in a hotel)"? (I know you'd prefer some additional information. If we're talking about a towel used by a male supermodel at a Four Seasons, rather than one wiping off a drug-addled, unhygienic pimp at a hovel at a Motel 6, that could make a big difference. *But* we're playing the all-else-being-equal game here.) Interestingly,

ment, our potential for disgust has only been dulled, not put into a coma where absolutely nothing about our partners' bodies can shake us out of our erotic trance. In fact, the system evolved with its own sensible limits. Surely you'd be excited to finally tear off the clothes of that attractive coworker you've long had your eyes on, for example, but removing his underwear and observing what appears and smells to be a raft of expired, sizzling bacon strips would put a damper on things. Or here's a homework assignment. The next time you're online and find yourself—funny how that happens—in a lascivious state while masturbating to your favorite porn, abruptly switch gears to do a Google image search for sexually transmitted infections. If you're an ovulating female, be forewarned: coming so suddenly like that upon the monstrously deformed and scab-covered penises of the afflicted may well make your ovum frantically double back into your fallopian tubes in a reversed race to the mother ship. And straight males blinking their eyes at all those multicolored vulvae with eruptions of syphilis, genital warts, crabs, gonorrhea, scabies, pelvic inflammatory disease, herpes, or any other weeping, oozing, bursting incarnations of a true libidinal nightmare can expect to undergo penile detumescence at warp speed. (I think I'm finally starting to get what that foot fetishist saw in me. Against the backdrop of festering genitalia, my feet are quite handsome, really.)

What this sojourn into such a vivid venereal hell tells us is that if the disgust gets really potent, our brains have an emergency "kill switch" that can shut down the whole lustful operation no matter how far things have already gotten under way. Seeing such

---

women with a history of vaginismus (the difficulty in allowing vaginal entry to a penis, a finger, or other object despite wanting to be able to do so with their partners) are significantly more disgusted at the prospect of touching these soiled objects than are control subjects. "From this perspective," says Jong, "the difficulty in penetration in women [with] vaginismus may at least partly be due to a disgust-induced defensive response." Another way to think about this is that such women are resistant to the local anesthetic. See Peter J. de Jong et al., "Disgust and Contamination Sensitivity in Vaginismus and Dyspareunia," *Archives of Sexual Behavior* 38, no. 2 (2009): 244–52.

clear signs—blinking, throbbing marquees, effectively—of a communicable sexual disease and going forward with the sex act anyway wouldn't have been a very adaptive evolutionary decision for our ancestors. And as a safety precaution, our sex drive is also likely to remain out of service for a while in the aftermath of such a traumatic sensory affair. After all, the brain has now received vital information about venomous genitalia being out there in the social environment, and some (the asymptomatic cases) might not be so painfully obvious.

The overall effect is similar to what happens with a food-related conditioned taste aversion. Whether it's an accident of timing and menu choice (such as a chance coupling of the flu and potato salad), or a bad bout of nausea or hurling our guts out after we've incorporated some actual contaminants that have made us sick (perhaps a disgruntled cafeteria worker laced that all-American picnic staple with arsenic), our bodies are exquisitely designed for one-shot learning in avoiding hazardous foods. Once you ingest that particular food and soon after get violently ill, you don't have to convince yourself to avoid it tomorrow. In the present case, one day down the road your body may let you revisit potato salad as a possibility, but for the foreseeable future your stomach rejects it as a viable option.

Likewise, the feeling of overwhelming disgust while we're sexually aroused can cause us to avoid for a while whatever erotic situation it was that put such an instant crimp in our libido. We might call this the "conditioned carnal aversion" effect. In the past, this mechanism was, sadly, exploited by psychiatrists attempting to "cure" gays and lesbians. In a behavioral modification technique known as "covert sensitization" (a form of classical conditioning), patients are told to visualize a highly unpleasant scenario occurring in tandem with their unwanted behaviors or desires. If you were an unhappily gay male patient of the psychiatrist Barry Maletzky in 1973, for instance, you might have been given the following story to help rid you of your cravings for other men's privates.* "You're at the beach with a special person, John," the vignette began:

> Imagine the ocean, the smell of salt air. You lie down be-
> hind a sand dune and start to embrace and undress each
> other. You can see his penis hard and stiff. He starts rub-
> bing it back and forth. But as you get closer you notice a
> strange odor and you see small white worms, like lice,
> crawling in the hair around his penis! You're touching
> them with your mouth! It's disgusting and it's making you
> sick. Some of them have gotten onto you. They're crawling
> into your mouth. Your stomach starts to churn and food
> particles catch in your throat. Big chunks of vomit come
> into your mouth. Vomit dribbles down your chin. You can
> see the worms still, crawling in your puke, and you get
> sicker than before.

Such an intervention didn't exactly turn gay men straight, but it did occasionally turn them off from having sex with other men for a while. (After reading an equally disgusting story featuring a worm-eaten vagina, after all, I'm guessing a lot of straight men would probably avoid those for a time too.) Since the effect would wear off before long and the man's lustful gay desires eventually return, some clinicians recommended that the pa-tient keep a handy flask in his pocket containing a foul-smelling liquid or chemical. Armed like this with his very own portable emetic, the patient could then quickly pull out the bottle and hold it under his nose to invoke a vicious bout of nausea (better yet, vomiting) whenever he became "inappropriately" aroused by the same sex. It's kind of like those poor copulating rats being

*This was right around the time that the American Psychiatric Association removed homosexuality as a mental illness from its official diagnostic manual (the Bible-like *Diagnostic and Statistical Manual of Mental Disorders*). Various methods of conver-sion therapy, however, remained common for years to come in private clinical practices.

given a shot of lithium chloride to turn them asexual, only here some gay people had learned to hate themselves as "perverts" so much that they could even be taught to self-administer their own poison.

❧

In the politically less dramatic world of heterosexual desire, there are interesting mechanisms in their own right when it comes to lust and disgust. For example, there's reason to believe that the precise hydraulics involved in these counterbalancing forces evolved to work in slightly different ways in the brains of men and women. The best way to interpret the sex differences we're about to examine is through the adaptive lens of "parental investment theory." There are many complex layers of caveat and detail to the biologist Robert Trivers's classic evolutionary theory, but its basic ideas (and those now accepted by experts as truisms) have to do with the costs of casual sex being unequal for males and females. (And note that the theory clarifies that our sexual brains work the way they do *today* as the result of conditions faced by our ancestors *tens of thousands of years ago* when they first evolved—a time when birth control and prophylactics weren't anywhere to be had.)

Women are born with a finite number of eggs (one to two million on average) and can bear only a finite number of offspring before reaching menopause. Pregnancy, childbirth, and breastfeeding all place tremendous demands on the female body. Furthermore, nine long months of gestation, several of which can make women largely dependent on others for their survival, are followed by a period of infertility while lactating that puts them even further out of reproductive commission. For a woman in the ancestral past to have sex (thereby placing herself at the risk of conceiving) was therefore to have assumed a significant "cost," one that even included the possibility of her own death through pregnancy- or birth-related complications. Under these conditions, natural selection favored a conservative investment strategy for her ovarian assets, which meant women were generally wary of casual sex and relatively choosy about their sexual partners. (These are gender *trends*, needless to say, not hard-and-fast *rules*. We'll also be encountering plenty of folks whose sexualities stray far from these normative patterns.)

Men's bodies, on the other hand, produce around eighty-five million sperm cells per day, per testicle. Over the course of an average life span, that adds up to more than a quadrillion gametes. With virtually unlimited holdings like these—and being virile on any given day of the month, too—a bit of seminal splurging here or there wasn't "costly" at all for our male ancestors. Rather, evolutionarily speaking, sex is cheap for men. This is especially true, of course, since it's not the male who could end up with an adorable but demanding little incubus inside his body cavity growing exponentially for the better part of a year only to be, at long last, expelled painfully out of his urethra. Instead, if his female partner got pregnant as the result of his three-minute time sink (the average latency for a man to ejaculate into a vagina), then all the better for his genetic replication. Given these factors of having gametes galore and being immune to pregnancy, the most adaptive investment strategy for human males, so the Trivers theory goes, involved a general penchant for casual sex and not being too picky about one's sexual partners.

With these differential costs in parental investment between the sexes, then, the key for each would have been to experience sexual disgust to the *right degree* and by the *right stimuli*. Let's start with the bells and whistles of the lust-and-disgust hydraulic system at work in the evolved male brain. Bearing those earlier investment strategies in mind, note that any man who was too choosy, too often, and too easily grossed out would have frequently missed opportunities to spread his genes. With spermatozoa by the mega-millions chomping at the bit in his gonads and trillions more to take their place, easily *overcoming* sexual disgust, especially when it came to having sex with "receptive" females, would have been a pressing adaptive problem for the ancestral human male.

And by the overpopulated state of things today, he seems to have solved this problem like a rock star. You'll recall what Freud said about the human libido and how it "thrives on obstacles." Once a strong degree of male arousal occurs, only the most blatant of maladaptive emergencies can trigger the erotic kill switch. Otherwise, the male gene-reproducing operation is lustfully motivated to go forward at all costs. This even applies to instances for which, in a less aroused frame of mind, the male would clearly see the massive long-term risks he's taking on by proceeding with such a reckless romantic mission.

When it comes to sexually transmitted infections, for example, more than a few dangerous viruses (both those of today and those of the past) could again be asymptomatic. Signs of overt disease will loyally deflate an erection into a protective state of flaccidity, and this aversion may even be temporarily generalized to other women for the reasons we've gone over. But assuming that no such conditioned carnal aversion has occurred, *any* unprotected sex (which, remember, was the only form of sex our ancestors were having) has always come with a health risk for men. From an unconscious evolutionary perspective, it was a risk worth taking. That's according to Richard Stevenson, anyway, the researcher who had those male students fondle lubricated condoms and cold pea-and-ham soup. He reasons that "[men's] need to 'lower the guard' can be seen as one of a large number of balancing acts that organisms must engage in to optimize breeding success."

In other words, over many human generations, men who underestimated the threat of disease from casual sex outnumbered those who were duly cautious about their chances of getting infected. Natural selection plays the numbers game. And overall, more men would have survived impulsive sexual decisions than not, thereby replicating their genes. So those old rubs about "thinking with the wrong head" do, in fact, accurately reflect the crude, stereotypically impassioned male, but there's evolutionary logic to such idiocy. Many of us alive today, men *and* women, are the descendants of males whose heedless passions effectively short-circuited their long-term reasoning abilities.

A study by the psychologists Hart Blanton and Meg Gerrard is especially telling in this regard. These authors asked a group of straight male undergrads attending a large Midwest university to rate the likelihood of their contracting HIV from having unprotected sex with one of nine hypothetical women. The participants had only two bits of information to go on. First, they were informed about the total number of men that each of these women had already slept with (either one, three, or eight previous male partners). Second, they learned about this hypothetical woman's use of condoms in the past (either "extremely good," "pretty good," or "not very good" about using protection).

Men who were presented with just the black-and-white facts about these fictitious females came across as responsible and mature. They concluded, logically, that their risk of acquiring HIV from having unprotected sex with the hypothetical women would increase as the number of previous amours stacked up and condom use declined. By contrast, those men who were presented with these very same biographical facts, but who *also* had these facts paired with images of attractive female models and were instructed to first fantasize about having sex with the ladies depicted, almost completely ignored all those "boring" details of the characters' past sex lives. Since the provocative image of her had managed to get them sexually excited, in other words, these men saw only minimal HIV risk in having unprotected sex even with the woman who had the most prior partners and the "not very good" history of condom use.

Men who are in such an intoxicated state of desire aren't only liable to make (really) bad decisions that put their health on the line but also more prone to making poor decisions that can have serious long-term consequences for their social lives. Lust is a sort of mental fog in which men temporarily lose, or at least lower, their standards. And by "standards," I mean in the sense of whom (or what) a highly aroused man is willing to have sex with as well as the moral values that he'll place into temporary quarantine just to satisfy his pressing urges. Again, from an evolutionary perspective in which sex for men was so cheap, that was indeed the best adaptive bet. But a bet also meant a gamble, and just as lust can undermine effective risk assessment in acquiring sex-borne diseases, this biologically adaptive mechanism of lust relaxing a man's standards in these ways came with the old gambler's proviso that there's always the chance one could lose everything.*

Examples of such bad endings, of passionate encounters hastily born and ending up in elaborate murder plots, suicides, sex-offender registries, lengthy prison sentences, complex love triangles, or a disgruntled fling breaking into your house and boiling your child's pet rabbit on your kitchen stove (that's a pop-culture reference from way back in the 1980s, for any confused youngsters out there), abound in human cultures. In the previous chapter, we saw how the concupiscent brains of Genet's town officials in *The Balcony* occupied a very different social space from that of their frigid alter egos. The bizarre goings-on at the madam Irma's whorehouse, however, were just playful romps compared with the scenes unfolding in Georges Bataille's cult novella, *Story of the Eye*. More slasher flick than love story, *Story of the Eye* is a tale about two teenage sexual psychopaths. "That was the period when Simone developed a mania for breaking eggs

---

*Note also that the odds fall decidedly against the lustful male "gambler" if he has a sexual fetish or paraphilia deemed criminal or socially inappropriate. The underlying psychology of lust serving to lower moral standards is the same for all men, but such disinhibition places the deviant at a distinct disadvantage in this sense over other males.

with her ass," the now-adult narrator recalls fondly of his sweet-heart. (The book was first published in 1928, so this was breath-takingly scandalous in its day.)

The penultimate scene in *Story of the Eye* involves the lewd and lascivious pair luring an unsuspecting priest into a decadent violation of his celibacy vows. Once Simone works him up into an unholy state in the confessional booth, the highly aroused priest—an honorable man of faith corrupted by this bewitching adolescent temptress—puts his sterling morals on the back burner for a moment. Bataille writes vividly: "His body erect, and yelling like a pig being slaughtered, [the priest] spurted his come on the host in the ciborium, which Simone held in front of him while jerking him off." In *The Balcony*, Irma observed how the mental fog that caused the town's male officials to be per-verts clears up instantly once the sex act is done. Bataille's priest in *Story of the Eye* felt this same instantaneous impact of moral clarity. Only it hit him a little harder. "Now that his balls were drained," Bataille tells us, "his abomination appeared to him in all its horror." (It's just downhill for the priest from there, too. He's immediately strangled, gets one of his eyes gouged out, and then Simone even puts it inside her . . . well, I'll stop there. But hence the title of Bataille's book.) It's certainly on the extreme side, but Bataille's *Story of the Eye* is the archetypal tale of a good man being led over the edge by his own irrepressible lust.

Nearly a century later, the social psychologists Dan Ariely and George Loewenstein confirmed this same psychological ef-fect in a controlled experiment. Fortunately, their 2006 study included not a jittery priest, teenage sexual psychopaths (at least not that we know of), or even a ciborium but thirty-five of your average straight male undergrads from UC Berkeley. It was a straightforward experiment. Simple, really. Roughly half of the participants (the control subjects) completed a questionnaire at home about a variety of sex acts, or more specifically about whether they could ever see themselves partaking in the act under consideration. These were the types of questions you might overhear in a school cafeteria being bandied about as philosophical quandaries by a bunch of eighth-grade boys. "Could it be fun to have sex with someone who was extremely fat?" for example, or "Can you imagine getting sexually excited by an animal?" "Can you imagine having sex with a 60-year-old woman?" and "If you were attracted to a woman and she proposed a threesome with a man, would you do it?" You get the idea. These were sex acts, in other words, that only a select group of deviants would claim as appealing. And, alas, no such individuals could be found in the control group of Ariely and Loewenstein's study. Rather, as you'd expect, these college-aged males expressed disgust and disdain for such unpopular bacchanalian festivities.

Remember, however, that only half of the participants had been randomly assigned to the control condition. The rest were given a more hands-on assignment. Prior to answering the very same questions, these other men were instructed to masturbate to their favorite porn at home—get as excited as you can, they were basically told, but do kindly refrain from climaxing. Once they got to the highest possible summit of this needful, aching state of arousal, they turned their attention to the items in the questionnaire. Interestingly, their responses turned out markedly different from those of the controls. These lustful men were much more "open-minded," not only for the examples from before, but also for sadomasochism, fetishism (shoes, sweat, cigarettes), rape, and pedophilia. Our old friend Havelock Ellis would have been especially delighted about their answering affirmatively, proudly even, to the token urophilic question, "Would it be fun to watch an attractive woman urinating?" Most men, as these data show, are only a lustful cogitation or two away from morphing into "real perverts" who "go against what is right."

•

When it comes to human evolution, men's sexual arousal is of course only half of the equation. Women also appear to come equipped with their own specialized arsenal of disgust-related adaptations, ones similarly designed in the female's best reproductive interests. Given the comparatively higher parental investment cost of casual sex for a woman in the ancestral past, for instance, one might expect female sexual arousal to actually ramp up, or at least to *strengthen*, her moral resolve. After all, whereas the main adaptive problem for men involved diluting their disgust in order to be able to have fruitful sex with even the most unpalatable of prospects (that is, lowering their standards), women would have instead benefited from an even more sharpened sense of disgust because it navigated them away from undesirable reproductive partners. A man who, intentionally or otherwise, happens to impregnate some poor random woman through casual sex one particular day (say, a specimen from Henry Miller's *Tropic of Cancer*, one of those spicy Parisian ladies with an "aggravated misfortune" such as "a missing tooth or a nose eaten away") can just impregnate a more genetically fit one the next day. If not, he can always try again the day after that (or perhaps later that evening). By contrast, a woman who has just been impregnated by an undesirable man (quite possibly that very same undesirable whom we've just met) can't move on to better mating prospects so quickly. It will be a considerable period of time before she can try to get pregnant again—and with a baby in tow. With all of those imposing biological costs entailed for women, a comparative pickiness in sex partners became paramount to female genetic success. And sexual disgust for male undesirables in all their misshapen forms—morally, socially, and physically—was instrumental to solving this uniquely female adaptive problem.

Results from a study by the anthropologist Daniel Fessler and his colleague David Navarrete support the notion that women's disgust for "biologically suboptimal unions" flares up especially when they're ovulating. These researchers asked hundreds of reproductive-aged women (who, naturally, were at various phases of their menstrual cycles at the time) to fill out an online questionnaire concerning a broad range of hypothetical hookups. Many of these hypotheticals closely resembled those earlier questions from Ariely and Loewenstein's study of disgust in aroused men. Fessler and Navarrete's list included, for example, "a 20-year-old woman who seeks sexual relationships with 80-year-old men" and "an adult woman who has sex with her father." Their hypothesis was that the female participants' disgust at the very thought of such "biologically suboptimal" pairings would vary as a function of their state of fertility, reflecting a psychological adaptation that helps to discourage women from making maladaptive decisions when the stakes are greatest. Indeed, those ladies who were "high in conception risk" while responding to the questionnaire were significantly more repulsed by deviant couplings involving bestiality, extreme age-disparate unions, and incest than those at less fertile periods of their cycles. Furthermore, just as we'd expect with our local anesthetic model from earlier, these women were no more or less disgusted than their not-quite-as-fertile cohorts when it came to items featuring more general yuckiness (such as thoughts of maggots on a piece of meat, picking up a dead cat, or stepping on an earthworm with their bare feet). Rather, it was only the cases of sexual deviancy that made these very fertile women's disgust ratings really stand out.

Fessler and Navarrete's study wasn't a "within-subjects design" (which means they didn't check to see if the disgust ratings of the same women fluctuated over the course of their own menstrual cycles). However, with a randomly drawn "between-subjects design" (which means they compared the disgust ratings of different women as a function of their present fertility status), there's every reason to assume that these cyclic effects in female disgust sensitivity also apply at the "intra-individual" menstrual level. In other words, when she's more likely to become pregnant, and therefore poised to make critical mating decisions that are either biologically *adaptive* (such as sex with a healthy, successful, similarly aged partner who'll help to support the child), *maladaptive* (such as sex with an octogenarian who might not even survive the length of her pregnancy), or simply a *poor use of her time* (such as sex with a sea turtle), the average woman becomes significantly less "open-minded" about sexual deviancy than she may otherwise be.* But assuming that her male partner is human, and given that his genitals aren't crumbling away from disease, he's not any older than her great-grandfather, and he's more or less a reasonably decent mate, the lustful mind-set of a fertile woman encourages her to have productive intercourse, with that reliable disgust anesthetic making it all possible.

So far, we've been talking about disgust in its more literal forms: as a contamination-avoidance mechanism with an elaborate pulley-and-lever system that evolved to help us make biologically adaptive decisions in the heat of the moment. But disgust has also come to have powerful *symbolic* elements, and these, too, are meaningfully tied to human sexuality. The often-dramatized, heartbreaking image of a woman crouched in the corner of a shower and frantically trying to scrub her body clean after being raped is indeed supported by empirical evidence. Seventy percent of female sexual assault victims report a strong impulse to wash afterward, and a quarter of these are still washing excessively for up to three months later. For women, simply imagining an unwanted sexual advance can serve to turn on this "moral cleansing" effect. In one study, two groups of female participants were told to close their eyes and picture being kissed. The members of one group were instructed to imagine being aggressively

---

*Fessler and Navarrete's data deal with disgust and not value judgments per se, but the findings hint that women's moral dumbfounding ("It's wrong *because* it's gross") may also peak when they're ovulating.

cornered and kissed against their will by an undesirable male. The members of the other group, by contrast, were asked to envision themselves making out with an attractive man in a consensual embrace. Only those women who'd been randomly assigned to the coercive condition chose to wash up after the study.*

In many cases, it's as though the person's very sense of self has been contaminated as the result of being sexually assaulted. Here's one young woman, for instance, describing the emotional aftermath of her childhood molestation:

> I could stand and stare at myself in the mirror and just want to be sick, I couldn't make sense of it, I couldn't understand that it was me it had happened to. And it was so bloody disgusting that I felt as though I was going to be sick . . . I stood in a bathroom and looked at myself in a mirror and tried to understand that it was me, it was so revolting, I think it ended with me sort of, I don't know I just switched off. I felt like the filthiest, most disgusting child in the world. It was really disgust, *disgust beyond description*. (Italics added)

When symbolic disgust gets into one's core identity like this, the psychological sanitation process is never an easy one. There's now a residual grime on the person's subjective filter through which she perceives herself, and if left untreated, these effects

---

*It works the other way around too, whereby people try to symbolically decontaminate themselves from the stain of their own lust. One of the most disturbing examples of this "moral cleansing" effect is the phenomenon of baby rape in post-apartheid South Africa. A widespread "virgin myth" among some men in that society held that the only way to cure an infected man of HIV was to have sex with a virgin. This led to some men seeking out the most "virginal" and morally pure sex partners possible (meaning, sadly, infants and toddlers). "Not only was the child violated . . . [by] the unmitigated and undiluted brutality of the perpetrator," writes the sociologist Deborah Posel, "but the risk of transmission of the HIV virus doomed the child to the prospect of death." Deborah Posel, "The Scandal of Manhood: 'Baby Rape' and the Politicization of Sexual Violence in Post-Apartheid South Africa," *Culture, Health, and Sexuality* 7, no. 3 (2005): 246.

can permanently darken and sully the individual's entire sense of being. The insidious consequences of disgust in generating feelings of hatred and loathing *of others* that we saw earlier (such as in the political ploy of fanning hatred for gay men by a rhetorical emphasis on anal sex) typically lead to a behavioral avoidance of the object of one's social distaste. In fact, the measurable physical distance placed between oneself and the hated target (such as in an elevator) can show this effect empirically. No matter which way our own worldview happens to tilt, we usually don't stand too close to people whom we believe harbor opinions or attitudes that are morally repellent to us, nor do we seek to place ourselves in the immediate vicinity of those who've engaged in social behaviors we strongly believe are offensive and wrong. Avoiding such a morally "disgusting" person gets far more complicated, however, when the primary source of your symbolic disgust is *you*. After all, there are only three ways to escape the self—depressive sleep, drugs, and suicide. And none of these, needless to say, is healthy.

Once a person feels tainted this way by an act judged to be especially unacceptable by his or her own society (either as the victim of the act *or* as the offender who feels genuine shame and remorse after his lust got the better of him), these rankling feelings of symbolic disgust can quickly metastasize into malignant self-hatred. Sexually abused children, for example, are far more likely than their peers to develop an exhaustive suite of psychopathologies later in life. Suicide rates skyrocket, and correlations have been found with everything from chronic depression to self-harm (such as cutting), substance abuse, eating disorders, paranoia, hostility, and psychoticism.

The most common way of managing the damage is to channel the harmful, caustic emotions elsewhere. Usually, this involves directing the symbolic disgust outward—away from the self—and toward those perceived to be responsible for sullying the self. A study by the psychologist George Bonanno, for instance, showed that the coping strategies of adults who'd been sexually abused as children could be reliably gauged by unobtrusively observing their nonverbal facial displays during a therapy session. Those who, as kids, hadn't disclosed their sexual abuse to others (for example, it was discovered by another adult and only then reported) and who blamed themselves displayed far more "non-Duchenne" (or fake) smiles than did those survivors who blamed their abusers. This latter group was more clearly identifiable by their facial expressions of disgust—a palpable *moral loathing*—whenever speaking about those who'd harmed them.

Although such powerful symbolic disgust responses are all too real in the damage they can do to a person's well-being, you may be surprised to learn that their precise parameters have no basis in a moral reality. Unlike food-borne or disease-ridden pathogens that human beings have evolved to combat through adaptive responses and that require absolutely *no* enculturation (we don't have to "learn" how to get diarrhea or vomit, for instance, after wolfing down a burger infected with *E. coli*, nor do we require lessons on how to go about experiencing a precipitous drop in desire on seeing—or smelling—our sex partner's worrisomely encrusted genitalia), what induces the symbolic disgust response is defined by prevailing cultural forces. That's to say, for the most part we've *learned* to morally loathe whatever it is that we've come to do; this information isn't genetically inborn but socially acquired. What might have made a Japanese person commit ritual suicide in the eighteenth century because he couldn't stand to live with himself and his shameful social offense would for most of us today be quickly forgotten as a trifling incident. Given their sheer emotional intensity, it's easy to mistake feelings of symbolic disgust for an immovable moral reality that exists outside our own subjective heads. But they don't.

Anthropologists have long known just how easy it is to make Western moral compasses spin out of control by describing other "exotic" cultural traditions, especially those involving sex. Consider the elaborate semen-ingestion ritual of the Sambia of Papua

New Guinea, in which around their eighth or ninth birthday all of the boys in the community are banished to a bachelor's hut filled with older males, and for several years thereafter they fellate these more senior figures on a daily basis. The Sambia believe that semen is a magical substance that transforms their youths into mighty soldiers, and the more seminal fluid that boys swallow, the more powerful they become. "By the age of 11 to 12," explains the anthropologist Gilbert Herdt, "boys have become aggressive fellators who actively pursue semen to masculinize their bodies." In our society, this "semen-ingestion ritual" would be *unspeakable*, causing irreparable harm and condemning these boys to lifelong issues with their sexuality. By contrast, Sambia adults and older teenagers who "donate" their semen to young boys are seen as altruistic. And within that society, the boys grow up to be well-adjusted adults who, in turn, give their own magical semen to the next generation. The Sambia perceive harm in *denying* boys participation in the semen-ingestion ritual, since this is to permanently brand these children as weaklings who were judged unworthy of defending the community as adult warriors.

Or consider, while we're on the subject of semen, the case of a Pennsylvania man accused of using syringes to inject what I once heard referred to as "baby batter" through the tinfoil lids of his coworkers' yogurt containers. Rather disgusting, you'll certainly say (and to which, I hasten to add, I completely agree). The judge hearing the case called it the "most despicable act" he'd ever seen—this with a long bench history overseeing rape, child abuse, and murder trials—and promptly sentenced the man to two years in federal prison. This man's actions might strike us as obviously antisocial, raising public health concerns as well as being a rather grotesque form of sexual harassment in the workplace. But there is a cultural context to every crime, even when it comes to tampering with a person's food by covertly lacing it with seminal fluid. If this troubled man had been a bachelor living in the Egyptian oasis of Siwa in the fairly recent past, a local sage there (the equivalent of our judge here) might have actually instructed him that the surest way to a girl's heart was concealing his seed in her favorite food. She, in turn, wouldn't be disgusted on discovering her suitor had done this, but more likely flattered.

The notion of abnormal sexuality is as much a matter of straying from our culture's sexual scripts as it is one of violating the laws of reproductive biology. For men, at least, being in a lustful state makes us more likely to wander over both of these lines, and sometimes dangerously so, hurting both ourselves and others in the process. But the concept of the "perversions" (or "going against what is right") is entirely a phantom of the moralizing human mind. When unburdened of its massive emotional weight, sexual deviance is no more and no less than a statistical concept that signifies being off course from our societal norms. Very little is universal when it comes to human sexuality. And once we acknowledge this lack of universality, the illusion that there's anything like an objective right and wrong in the vast domain of our species's libidinal relations shatters beyond repair. The best that psychiatrists can hope to do is to describe and treat sexual deviance in their own idiosyncratic cultures.

Oddly enough, a healthy dose of moral nihilism is the antidote for so many of the social ills connected to human sexuality. To adopt the most clear-sighted stance on these increasingly slippery subjects, we've got to remember to take "deviance" within its given context, and "harm" must be understood as harm experienced by the parties involved, not by us as "disgusted" onlookers. Morality isn't "out there" in the world; it's a *way of seeing*, and it's constantly evolving. In fact, as we're about to learn, we don't even need to look to exotic others to put our critical-thinking abilities to the test in this way. The emotional atmosphere of our own culture has undergone such radical social climate changes that to assume we're now finally glimpsing a clear moral reality that previous generations simply didn't notice because of their ignorance and cloudy biases would be stupendously foolish of us.

THREE

SISTER NYMPH AND BROTHER SATYR

People in general let loose the tiger of sexual desire they
have kept under leash and occasionally ride on its back until
they tumble into the Valley of Ruin.
                    —Ogai Mori, *Vita Sexualis* (1946)

In the days when there were still kings in France, there was
King Louis XIII, or Louis the Chaste. On the throne from 1610
to 1643, the least erotically minded monarch in French history
was, it turns out, a pretty active homosexual. I'm referring not to
the curious detail of his setting an international fashion craze for
men's wigs that lasted centuries but to the fact that he spent
much of his reign in bed with a crimson-headed marquis who'd
been gifted to him as a sexual companion by his sensible first
minister. Although his own "chaste" glands were actually getting
sufficient venting all the while, the sexual desires of Louis XIII's
female subjects were severely stifled. In fact, even an expression
as innocent as that—"sexual desires"—if used in reference to
female arousal would be something heard only in the somber
context of a serious medical consultation. Back then, diagnosing
a woman with a libido transpired in the way that doctors today
stare down nervously at the floor while breaking the news to

family members that their loved one has, indeed, tragically lost *both* arms to the explosion at the factory.

One such seventeenth-century Frenchman was Lazare Rivière, the eminent physician-scholar of Montpellier during the reign of Louis XIII. Rivière was a specialist, a pioneer you might even say, in the clinical treatment of female lust. It seemed this disease really was an epidemic throughout France. Why, in Montpellier alone there was a rash of lonely old widows stalking the picturesque meadows in pursuit of penises to sate their venereal appetites, and young girls said to be doing unthinkable things to themselves with (brace yourself) their own hands. That's not even to mention all the aging lady virgins still without suitors, nor the many women who *had* married but found themselves now with impotent old husbands. Just like their single contemporaries whom the heavens had so cruelly burdened with ravenous internal genitalia, these women's needful loins couldn't be satisfied by an amorous spouse's purposeful appendage, either. Rivière saw all such creatures as equally dangerous (after all, their desires might lead them to seduce a married man, and *that* sort of thing was especially frowned upon by His Majesty, King Louis the Chaste). But Rivière also felt that these women should be pitied and cared for. He insisted that many of them were chronic sufferers of a terrible mental illness, one that he liked to refer to as the "madness from the womb." That's to say, they were *sick*.

According to Dr. Rivière, the central problem was that noxious gases emanate from a piling up of the female's unspent "seed." Like fuliginous tentacles with the worst of intentions, these fumes would then creep up from her midsection and into her nervous system, where they'd interfere with her ability to think clearly. Even a normally pious and reserved woman could go insane with passion this way. "It's the abundance of seed . . . the parts made for generation, are vehemently stirred up, and inflamed with lustful desires," wrote Rivière. One can almost picture him patiently explaining this to a local farmer fretting over his growing daughter and the girl's unmentionable deeds with the carpenter's son. Flipping through the vellum pages of a hefty book that was his *Praxis medica*, he'd land with a sure finger—*Ah, oui, je l'ai trouvé!*—on that all-important line that spelled it out for the man so clearly: "From the same seminal matters so affected, vapors ascend unto the brain which disturb the rational faculty, and depose it from its throne." Over time, Rivière decried, this process "turns into a true and proper madness."

So there you have it: the modern woman's guide to *madness from the womb*. But it wasn't all bad news, according to this French physician. There were a number of proven techniques for easing the suffering of women whose overheated uteruses caused them to "openly before all the world ask men to lie with them expressing the action of generation in the broadest language their mother tongue affords." Among these salves were a diet of bland meats (spices would only whip up the woman's lust), leeches applied to the labia, baths filled with cold lettuce heads, and abstaining from dancing and romance stories. No resting their heads on comfortable pillows, either. Such materials were far too sensuous. Eventually, marriage to a lusty young man would do the trick for those girls with enviable prospects. Such a conjugal stud would be the best solution of all for the devout Christians of Montpellier. In the meantime, or for women who were more likely to go permanently unpartnered, Rivière scribbled out the following prescription:

> The genital parts should be by a cunning midwife so handled and rubbed, as to cause an evacuation of the over-abounding [seed]. But that being a thing not so allowable, it may suffice whilst the patient is in the bath, to rub gently her belly on the region of the womb, not coming near the privy parts, that the lukewarm temper of the water

may moderate the hotness of the womb, and that it may by the moisture be so relaxed, as of its own accord to expel the seminal excrement, and that nothing else be done with the hand, save a little to open the womb, so as the water may pass into its more inward parts.

Peculiar as Rivière's ideas sound today, the medicalization of female arousal has historically been more the norm than the exception. Such an approach to women's lust certainly didn't end with the Enlightenment. The feminist sociologist Carol Groneman has traced the long and depressingly misogynistic history of reproductive medicine, focusing most of her attention on those infamously rigid Victorian attitudes of the nineteenth century. During this period, women continued to be seen as either passive receptacles for men's pleasures or chattel for the purposes of men's breeding. (Either way, pathological modesty dictated their flesh be drowned in heavy garments.) Just as in Rivière's day, expressions of female lust were taken to be signs of illness. The Victorian age is when this viewpoint went global, with doctors around the world now using the diagnosis of "nymphomania" for women with excessive desires (what constituted "excessive" was anyone's guess, but it usually meant any woman with a pulse).* Nymphomania was just a fancy new way of saying "madness from the womb," really, with just as much going for it as a scientific concept. Indeed, some gynecologists were recommending anti-nymphomania treatments strikingly reminiscent of Rivière's quick fixes.

For example, in 1856, a twenty-six-year-old doctor named

---

*Although the Victorian age is when the term "nymphomania" ("female disease characterized by morbid and uncontrollable desire") gained in popularity, coinage of the word is credited to the French physician D. T. Bienville in his 1775 thesis *Nymphomania; or, A Dissertation Concerning the Furor Uterinus*. In Greek mythology, nymphs were minor female deities who were usually depicted as nubile young maidens guided by their amorous passions and mating indiscreetly with mortals of both sexes.

Horatio Storer counseled a woman two years his junior who'd
made an appointment to see him on account of her very naughty
dreams. She also had a much older husband whose flagging erections, she confessed to Storer, couldn't satisfy her own insatiable
sex drive. The young doctor's advice to this patient makes one
wonder if he had an ancient dog-eared copy of *Praxis medica*
tucked away in his desk: stay away from meat, he admonished
her, no brandy, and replace the duvet feathers at once with
something less sumptuous, like horsehair. He did progress beyond cold-lettuce baths. None of that old-fashioned stuff for the
forward-thinking Storer; instead, he calmly instructed the young
woman to dab her vagina with borax solution to cool her inflamed passions. (At least it was better than the carbolic acid that
other doctors were recommending at the time.) But hers was
quite a serious case of nymphomania; it had reached such an
advanced stage that the poor thing had even taken to *masturbating*, of all things. "If she continued in her present habits of indul-
gence," Storer notes with colorless authority, "it would probably
become necessary to send her to an asylum." Years later, Storer
became president of the American Medical Association and one
of this country's first antiabortion crusaders.

Storer was quite a piece of work, but he wasn't alone in his
dyspeptic views of female sexuality. Alarmed by the specter of
loose women on the prowl, other Victorian-era gynecologists
warned their colleagues to be on the lookout for "seductresses"
fabricating symptoms of urine retention only to get unsuspecting male doctors to palpate their pudenda. Medical misogyny
even found its way into popular culture. British journalists explained how a nymphomaniac could be detected hiding among
"normal" females (a snake in the grass, in other words) by her
penchant for wearing perfume and flashy jewelry. Even speaking openly of marriage, it was thought, betrayed a woman's lewd
inclinations.

To complicate matters, physicians weren't on the same page

about what, exactly, nymphomania *was*. The diagnosis variably meant that the woman was having too much sex, that her desires were "clinically significant," or that she was a frequent masturbator. Also, given the social stigma that came with receiving such a diagnosis, gynecologists couldn't always count on their patients being honest about their sex lives. In some recalcitrant cases, doctors relied on "physical symptoms" of nymphomania to diagnose women with the condition. Genital hypertrophy was thought to be one such obvious clue to a woman's sickness. According to popular folk wisdom at the time, Mother Nature helpfully ratted out Messalinas by branding them with large clitorises, a trait noted as frequently occurring alongside these sickened, improper needs. (The misogyny behind such a "warning sign" is indefensible, but in fact there may be something to this clue. When testosterone treatment is prescribed for women today, side effects often include an enlargement and sensitivity of the clitoris—coincident with a rise in sex drive.)

These were dark days for women. Surgical clitoridectomies were even being recommended as a last line of defense against the great ill of female masturbation. One of the most infamous advocates of this barbaric procedure was an English gynecologist and obstetric surgeon named Isaac Baker Brown, who believed that everything from epilepsy to mania to catalepsy in women stemmed from their self-pleasuring habits. His professional undoing came when he performed the surgery on several patients without their consent while they were under general anesthesia for more routine procedures. Ironically, when Baker Brown died, the dissected brain of this practitioner who, under the cloak of Hippocrates, was so eager to dispose of these body parts of female pleasure was found riddled with syphilis courtesy of his own "excessive" needs.

It wasn't just the British who were such prudes during the

Victorian era. As we saw with Storer's handling of his patient, the concept of nymphomania had crossed the oceans. Here's another American woman of the period describing her battle with masturbation: "While I was praying my body was so contorted with the disease that I could not get away from it even while seeking God's help." A tad melodramatic to us today, but that was the mind-set. And it was indeed a warped way of seeing female sexuality. In 1894, an overwrought mother brought her nine-year-old daughter to the New Orleans physician A. J. Block after discovering the little girl masturbating. Block propped up the child on his examining-room table and began inspecting her genitals with his fingers. There was no reaction upon touching her labia. But "as soon as I reached the clitoris," the doctor later recounted without any emotion, "the legs were thrown widely open, the face became pale, the breathing short and rapid, the body twitched from excitement, slight groans came from the patient." In Block's imagination, these responses clearly meant that the child had a very bad case of nymphomania indeed. So, with the approval of the girl's mother, he performed an emergency clitoridectomy on her.

When we hear the phrase "female genital mutilation," our thoughts usually sail over to Africa, but the practice of eliminating a woman's capacity for sexual pleasure by removing critical parts of her anatomy has a distinctively Western history, too. The gynecologist John Studd (an improbable name given his profession, but true nonetheless) believes that more clitoridectomies were performed in England and the United States over the past two centuries than we'd care to recognize. One of the first uses of radiotherapy was the obliteration of teenage girls' clitorises to discourage them from masturbating. These X-ray clitoridectomies weren't happening in backwater clinics, either, but in some of the most fashionable cities in the world, including London and Manhattan. And this was just in the twentieth century. Fortunately, radiotherapy quickly moved on to its more benevolent purposes (such as the therapeutic treatment of cancer patients, a *slightly* more humane practice). But the fact that this shiny new technology got its start with such a cruel and unnecessary procedure should give us all pause. Most of these "patients," after all, were just healthy teenage girls whose parents couldn't bear the thought of their daughters doing *that*.

The clinical concept of nymphomania was still floating around as recently as 1964, when the book *Nymphomania: A Study of the Oversexed Woman* hit the shelves. It's tempting to dismiss a book with such a sexist title as hogwash—and much of it really *is* hogwash. But it was taken perfectly seriously at the time, mainly because the well-known psychotherapist Albert Ellis, the founder of cognitive behavioral therapy, was its author. In the book, Ellis (who was no relation to Havelock, by the way) introduces us to several different "types" of nymphomaniacs. Take the twenty-seven-year-old career woman "Dolores," the "Conquering Woman Type." To understand Dolores's nymphomania, the psychologist reasoned, one had to appreciate the self-consciousness she'd long been dealing with over her heavily scarred face, the result of a freak childhood accident. She "found herself," wrote Ellis, "with an extremely feminine body to go with her disfigured face." It seems that Dolores had gotten herself into the habit of making conquests of men and then promptly discarding them, with no interest in dawdling in relationships with those who fell in love with her along the way. From one to the next—virgin office boys, married businessmen, clerks long in the tooth and overdue for retirement—she delighted only in the numbers accumulated. Dolores wasn't a fan of foreplay either. "She wanted the whole thing or nothing," Ellis shares with his readers. It's hard to imagine the part in their interview where the woman explained to him her talents in inducing multiple orgasms in men, which the author also tells us about. But in any event, in Ellis's view, promiscuous sex pacified Dolores's self-consciousness, making her feel desirable despite her disfigurement. (With that psychoanalytic critique in mind, I can't help but feel it's probably for the best that Dolores didn't know *then* what we know *now* about the evolution of male arousal and disgust.)

Then there was Ellis's case of "Gail," or more precisely the case of "Gail and Burt." "Several homosexual-nymphomaniacal matings have come to my attention in recent years," Ellis begins his account of the "Neurotic Type" of the oversexed female. Such "matings" referred to those mismatched domestic partnerships, apparently quite common in the 1950s and '60s, in which a loose woman sets up a home with a gay man. The psychologist speculated that more often than not in such cases, the homosexual male represents a sort of masculine safe haven that an insecure nymphomaniac can depend on without having to fear that she'll be traded in for a more attractive woman.

Gail was a single mother, and Burt was helping to raise her young son. The pair was endowed with perfectly compatible personalities and reproductive anatomies, but given that both had an eye for men, never the twain shall their genitalia meet. "Now, let's not be so *cynical*," Ellis would surely have interjected in reaction to our pessimism. Nymphomaniacal Gail, you see, was also deeply in love with the flamboyant Burt. Not only was he attractive, he was also intelligent, a good conversationalist, and a caring father figure to her child, whose biological father was apparently not in the picture. Gail would happily give up her wild ways if only her gay BFF would cultivate a passion for vulvae. Namely *hers*. Wasn't there anything, anything at all, she could do to get Burt to burn with desire for her rather than for other men? "I told Gail that this was theoretically possible," Ellis explains:

since homosexuals are not born the way they are and can, and in some instances do, change remarkably, so that they can enjoy heterosexual activities. But I told her that Burt was not likely to change, because he did not consider himself disturbed. He insisted he thoroughly enjoyed his homosexual activities and showed no inclination whatever to come for therapy.*

This psychologist wasn't one to allow a little thing like patient consent stop him from trying to cure an inconveniently stubborn man of his homosexual ways, however. "If she wanted to try," Ellis tells us of his conversations with Gail, "there was nothing to be lost in attempting to seduce Burt into heterosexuality . . . [so] the two of us devised a plan of attack on Burt's heterosexual virginity." Basically, Ellis instructed his client to weasel her hands into Burt's pajama pants while he slept at night. Since the two lived together anyway, and since the house was so small that they had to share a bed (probably not the brightest idea, really), there was plenty of opportunity for such stealth maneuvers. Being in a dream-induced stupor would presumably minimize Burt's resistance to Gail's unbidden affections. *Technically*, Ellis prescribed a sexual assault, but let's not get lost in the semantics.

Ellis reports to us that Burt showed some "irritation" upon waking to discover the woman fondling him and rudely mouthing his thoroughly unimpressed organ. Eventually, however, with Gail hinting that she couldn't go on with a sexless relationship for much longer, the gay man began to shrug his shoulders and let the straight woman have her hopeless way. Then, one magical night a few months later, and to both of their surprise, Burt experienced an orgasm during one of Gail's persistent manipula-

*Whoever "Burt" was (or is), it's worth commenting on just how extraordinary was his refusal to defer to Ellis's psychiatric opinion for those times. To assert that one is not mentally ill, when practically all the world informs one otherwise, requires either an uncommon degree of self-delusion or an inhumanly defiant moral clarity. The ironic thing is that almost all experts today would unhesitatingly credit Burt with the latter.

tions. Slowly working their way up to some clumsily deliberate intercourse, similar transactions of viscous pleasure hesitantly followed. And soon the trio—Gail, Burt, and Albert Ellis—were cautiously celebrating Burt's newfound heterosexuality.

As it turns out, caution was warranted. It soon became apparent that Burt was only going along with this gay conversion plan because he was petrified of losing his otherwise pleasant home life, his close friendship with Gail, and his relationship with her young son, to whom he was very attached. After all, if a closeted zoophile can ejaculate into his wife by pretending that she's a horse, then an admirably motivated gay man can, at least on those occasions when his hormonal stars are perfectly aligned, also make out a subcutaneous Adonis in a woman's labial folds. Nonetheless, this admitted lack of sincere change in Burt made Ellis grow impatient with the unwitting gay client's failure to acknowledge his "disturbance." I can just picture the exasperated therapist snapping his pencil in two, taking a deep breath, and finally summarizing the trying case when he concludes: "He was doing the right thing for the wrong reason." The "right" reason to be straight, in Ellis's 1964 view, was self-evident: straight was simply what every person should be. The psychologist gives us no closure on the case of the "homosexual-nymphomaniacal" mating pair. But I'd like to think they're happily married senior citizens now, both with their very own husbands.

Compared with nymphomania, the medicalization of "excessive" male lust—at least when it comes to the heterosexual variety—is historically less noteworthy. But male genitals didn't go entirely unscathed during the moralistic heights of the Victorian era either. The next time you pour yourself a bowl of Special K cereal, think of the very special advice given to young wankers by the inventor of cornflakes, Dr. John Harvey Kellogg: "Boys, are you guilty of this terrible sin? Have you even once in this way yielded to the tempter's voice? Stop, consider, think of the awful results, repent, confess to God, reform. Another step in that direction and you may be lost, soul and body. You cannot dally with the tempter. You must escape now or never." And for parents at their wit's end with their masturbating sons, Kellogg offered this sage recommendation:

> [Circumcision] should be performed by a surgeon without administering an anesthetic, as the brief pain attending the operation will have a salutary effect upon the mind, especially if it be connected with the idea of punishment, as it may well be in some cases. The soreness which continues for several weeks interrupts the practice [of masturbation], and if it had not previously become too firmly fixed, it may be forgotten and not resumed.

If you were iffy about having your son's foreskin lopped off, Kellogg recommended it be stretched as tightly as possible over the glans (head) of his penis and sutured shut with a needle and wire to keep him from getting an erection. And if *that* didn't work, well, you could always lock up the boy's organ in one of the doctor's patented genital cages. For the young offenders themselves, Kellogg and other physicians of the period had plenty of stern warnings about the dire health consequences of masturbation. Tragic stories of boys going insane or blind, or of having deformed or mentally impaired offspring, were made readily available to prevent the horny male adolescent from engaging in this vice. "Such a victim literally dies by his own hand," wrote Kellogg in his wildly popular *Plain Facts for Old and Young* of 1888. Around the same time, G. Stanley Hall, the father of the field of adolescent psychology and the first president of the American Psychological Association, called masturbation "the scourge of the human race." Unlike Kellogg, at least he was willing to grant them the hallucinatory pleasure of their wet dreams—

he wasn't happy about it, mind you, but he regarded nocturnal emissions as being beyond the boy's control.

Such bizarre ideas about the harmless act of masturbation were tossed long ago into the waste bin of history. Fast-forward to modern Europe, for instance, where several nations—including the former capital of Victorian prudishness itself, the United Kingdom—now have official diktats stating that the experience of sexual orgasm is a basic human right. Adolescents of both sexes are encouraged to masturbate routinely to curb the transmission of STIs and to reduce teen pregnancies.

With the exception of onanism in adolescent boys, a subject for which physicians and scholars like Kellogg and Hall clearly had a thorn stuck in their paws, clinical descriptions of men whose animalistic desires couldn't be satisfied are conspicuously scant in the deep historical records. Women who demonstrated some arbitrary level of lust were usually victims of misguided, arrogant doctors (most of whom, of course, were men), whereas men who exhibited licentiousness were far more likely to be shuffled off to the penal system—perceived not as medical or academic curiosities but as criminals. Unlike Rivière's account of poisonous gases disrupting female sensibilities, there are no convoluted theories about the migratory patterns of frustrated testicles making old widowers woozy. (Ironically enough, we now know that lust *does* affect male cognition dramatically.) Neither will you find medieval prescriptions for treating an aroused man by rubbing his tummy while giving him a gentle hand job in a warm bath so that the water may lap against his prostate gland and relieve his seminal tensions. Lust has always been regarded as status quo for men—as a controllable vice, not a sickness. Unlike "nymphomania," in which female desire was perceived as anomalous, men who expressed their default lecher were rarely seen as mentally ill. Perverts, maybe, but not crazy.

There's one notable exception to this sexist historical divide between women who were sick and men who did sick things. This is the work of the psychiatrist Richard von Krafft-Ebing, and more specifically his influential study of sexual deviance that resulted in the 1886 publication of *Psychopathia Sexualis.* Krafft-Ebing believed that some men suffered from a mental condition called "satyriasis"—basically, nymphomania's male counterpart. Whereas mere masturbation was often enough to get a woman diagnosed with nymphomania, a man had to exhibit an extraordinary degree of carnality to receive the diagnosis of satyriasis. Satyrs weren't just your average players. These were men on the order of that depraved French nobleman Count Donatien Alphonse François de Sade, better known as the Marquis de Sade. And just like that eponymous father of sadism and its most infamous practitioner, Krafft-Ebing's male patients had come erotically undone (the marquis's favorite sex act was sodomizing a young girl while getting violently lashed by another man). The psychiatrist felt that these men couldn't control themselves due to their medical condition of satyriasis, a neurological disorder of an overwhelming sex drive that he strongly suspected was inherited.*

In *Psychopathia Sexualis,* Krafft-Ebing provides a few examples of men presumably battling this dreadful disease. One such person was "Clemence," a successful forty-five-year-old engineer with a familial history of psychiatric disturbances. One oppressively hot summer afternoon in 1874, Krafft-Ebing explains, Clemence was riding on a train bound for his home in Vienna, where his wife and child were eagerly awaiting his return after a long business trip. Suddenly he found himself getting so worked

*There's still no clear evidence that individual differences in sex drive are genetically determined, but it remains a viable hypothesis. A cluster of "hypersexuality" (the modern term for excessive sexual desire and equally problematic as a scientific construct, as we'll see shortly) was found among interrelated Hasidic Jews in Brooklyn in the late 1980s. See Nancy J. Needell and John C. Markowitz, "Hypersexual Behavior in Hasidic Jewish Inpatients," *Journal of Nervous and Mental Disease* 192, no. 3 (2004): 243–46.

up by the seat vibrations, the incessant prattle of the other pas-
sengers, and the roiling temperature that "he could no longer
hold out against his sexual excitement and the pressure of blood
in his abdomen." Frothing with lust, Clemence exited the train
at the nearest stop, which was the small town of Brück on the
German-Austrian border, about ninety-three miles southwest of
Vienna. Under the scorching sun, this monstrously aroused engineer dragged himself all over town in hope of finding a stray dog
(as in an actual canine; that's not a euphemism) to relieve his
agony discreetly in an alleyway. Failing to do this, and luckily so
for the other panting inhabitants of Brück that day, he stumbled
with fiery crotch in hand into the neighboring village of St. Ru-
precht. Here, in a polluted haze, the befuddled Clemence crossed
paths with an elderly woman who he thought might like to see
his erect penis. It turns out he was wrong about that. The old
lady screamed; he panicked and tried to embrace her and got
promptly pounced on by her neighbors, who held him to the
ground until the police arrived to arrest him on attempted rape
charges. "He said that he often suffered with such sexual excitement," notes Krafft-Ebing. "He did not deny his act, but excused
it as the result of disease." And, astonishingly, so did the judge
when Krafft-Ebing explained his medical theory of satyriasis to
him. All charges against Clemence were dismissed.

Psychopathia Sexualis also includes the curious case of
"Mr. X.," a man who'd come to the psychiatrist's attention in the
aftermath of a rather eventful wedding ceremony. Mr. X., Krafft-
Ebing tells us, was a rakish bachelor who'd finally decided to
settle down in matrimony after years of playing the field. Allow
me to set the scene for you. Picture, if you will, the affluent
Mr. X. attired in a fine suit and being escorted proudly down the
church aisle on the arm of his grinning brother. It's an idyllic image. Dust mites swirl in the sun-drenched rafters like a flock of
miniature angels. Family and friends stir eagerly in the pews; the
priest clears his throat in preparation for the vows he's delivered

a hundred times before; the organist hunkers down in melodious devotion; and the groom takes his place at the altar to await his bride, prim and coquettish in her spidery veil. But then the mood abruptly changes. Before his future wife is halfway down the aisle herself, Mr. X. turns to face the audience, unzips, and unleashes his priapic demon for all to see. It's unclear what happened next, but I'm sure you can fill in those gaps easily enough. Stranger things have happened, but I'm guessing the service didn't end with a kiss.

Krafft-Ebing collected dozens of stories like these of sex-crazed men. To him, satyriasis was a real disease that caused certain males to act out in inappropriate, and potentially harmful, ways. Yet unlike nymphomania, a hypothetical condition that captured the attention of nearly everyone, Krafft-Ebing's concept of satyriasis languished in both medical and academic obscurity for more than half a century. It's only in 1966, in fact, that the subject of excessive male lust as a mental illness makes its next earnest appearance, with the American psychotherapist Franklin Klaf's book *Satyriasis: A Study of Male Nymphomania* helping to give the long-forgotten issue a fresh elbow in the ribs.

Klaf had been troubled by the number of male patients appearing in his office who seemed predisposed to engage in self-defeating bouts of sexual gorging. Taking the old theoretical baton from Krafft-Ebing, he added a number of interesting claims of his own about satyriasis. For example, like other forms of psychotic breaks, he argued, the disease is characterized by a temporary disconnect from reality rather than a continuous mental state, and it's usually precipitated by a stressful event in the man's life. Klaf also concluded that "satyrs" were disproportionately attracted to underage teenage girls and as a consequence frequently faced legal problems, which is exactly how those featured in his book had ended up on his couch chatting with him about their problematic sex lives.

Many of Klaf's observations were quite insightful. His claim that upsetting life events can trigger bouts of satyriasis has found support in recent studies showing that a minority of men responds to feeling depressed by becoming *more* sexually active, not less. For a very long time, it was widely assumed that anhedonia (or depressed mood) goes hand in hand with a reduction in sex drive for both sexes, but it turns out that "hypersexual" men tend to respond to depression very differently, getting more easily aroused when they're depressed and seeking lust as a transitory escape from their negative feelings.

Nobody's perfect, though, and some of Klaf's other claims are indeed a bit suspect today. Take his theory about male-pattern baldness: "Satyrs display more than the 'normal' concern about baldness," he argued. "All men are somewhat self-conscious when it comes to receding hairlines, both front and rear. Most accept this hormonally determined phenomenon as part of the natural course. Not so with satyrs. They look for miraculous cures to wipe away the attrition of time, and they often fall prey to unscrupulous hucksters." I'm sure my mother would have agreed with Klaf on his baldness theory, since I vaguely recall my dad getting a perm during the Garfunkel era to plump up what hairs remained on his mutinous scalp, and this happened to be a period in my parents' relationship marked by some extramarital strains. Yet there's still no evidence that high anxiety over follicular fallout betrays a man's proclivity to go over the erotic edge.

Klaf was by no means the last scholar to try to confirm the existence of satyriasis. In 1995, for example, the psychologist Wayne Myers wrote a curious little case study about a man named "Alex." By this point in history, most psychologists had given up on trying to turn homosexuals into heterosexuals and were instead more interested in helping their gay patients adopt healthy sexual behaviors consistent with their own orientation. Alex was a gay case in point. In his early thirties at the time, he had cultivated a distinctive modus operandi in his tireless pursuit of sex on the mean streets of New York City. This involved carrying around a Polaroid image of his erect penis (these were the days before iPhones made penis-photo transport *so* much easier), an organ that he considered exceptionally large. There's no reason for us to doubt this, really. I guess he figured that his appendage was a work of art that ought to be shared with other men, not kept hidden away where nobody could appreciate it. So he'd wander about in public places flashing this photograph to attractive male strangers, especially those he suspected of having their own extra-large penises and who'd reward his generosity with a complementary erection.

"Successful encounters led to transient feelings of relief," writes Myers, "but unsuccessful ones where he could not produce an erection or the man would not examine his photograph led to tortures of the damned." Alex, we're told, typically had at least one sexual encounter a day over a ten-year period while cruising the bars and bathhouses of Manhattan. But his not-so-subtle method of seduction wasn't limited to gay-friendly establishments, and when you're displaying your penis to random strangers, this can cause problems with ambiguously oriented passersby (which is to say, straight men who didn't take too kindly to a photograph of his rigid member being shoved under their noses).

❧

When Myers was writing about Alex in 1995, most North American and British psychiatrists had already stopped using old-fashioned terms like "nymphomania" and "satyriasis" to describe those with an unusually active sex life. The more gender-neutral "hypersexual" is favored today, but the evolution of this terminology for excessive sex over the latter half of the twentieth century belies the messiness of the central construct. Some terms remained just as sexist as "madness from the womb" (for example, the "Messalina complex" or "Don Juanitaism"), whereas others conveyed a patently moralistic view or tone of judgment about the appropriate levels and forms of sexuality (such as "libertinism," "erotomania," "urethromania," "oversexuality," "compulsive promiscuity," and "pathologic multipartnerism"). There were even subtypes, such as "the frigid nymph," "the Casanova type," and "the sexual compensator." (Incidentally, being compared to Casanova may sound like a compliment, since history has a way of sanitizing its heroes, but this icon of chivalry had a decidedly less glowing reputation in his own day. Over his long and prolific career charting miles of untrammeled female flesh, Casanova also liked having sex with boys, cost several nuns their hallowed positions in the Church, was rumored to have nearly wedded a young woman who turned out to be his own estranged daughter, was robbed of a small fortune by a light-fingered prostitute, and is believed to have been infected with gonorrhea more often than not.)

When it comes to ascertaining sexual excess, modern clinicians have two main diagnostic guides at their disposal: the *International Classification of Diseases* (presently in its tenth version as the *ICD-10*), which is a publication of the World Health Organization and the primary reference for practitioners in Europe, Australia, and some Asian regions; and the *Diagnostic and Statistical Manual of Mental Disorders* (currently in its fifth incarnation, the *DSM-5*), which is the go-to source in the United States and Canada and the flagship book of the American Psychiatric Association. The *ICD-10* includes the diagnosis of "excessive sexual drive," and believe it or not, it's *still* subdivided by the antiquated gender terms of "satyriasis" (for males with excessive sex drives) and "nymphomania" (for females with excessive sex drives). By contrast, despite some recent attempts to add "hypersexual disorder" to the *DSM-5* as a genuine psychiatric illness, North American "hypersexuals" or "sex addicts" are today diagnosed under the umbrella label of "sexual disorder not otherwise specified."*

---

*The term "sex addiction" is contentious, and much of the debate over its use stems from sheer linguistics. Critics of this wording argue that the construct of addiction should apply only to a brain-based chemical dependency on endogenous physical substances—namely, alcohol and other drugs—and they point out that it's illogical to speak of "addiction" for evolved drives like sex. Saying that porn is "like a drug" works as a metaphor, in other words, but natural sensory experiences are not pharmaceuticals.

It was the psychiatrist Martin Kafka, a physician at Harvard's McLean Hospital, who led the charge in 2012 to get "hypersexual disorder" formally recognized by his peers as a real phenomenon with well-defined diagnostic criteria. Kafka's definition of hypersexuality was "excessive expressions of culturally tolerated heterosexual or homosexual behaviors"—for example, porn, random hookups, cybersex, strip clubs, erotic massage parlors, or masturbating so often that your exhausted genitals have recently started recoiling in fear at the sight of your hand.* In other words, Kafka's diagnosis was reserved for people engaging in "too much" socially permitted sex or having "excessive desire" for other consenting adults.† These he called "normophiles."

Yet Kafka narrowed down the criteria for his proposed diagnosis even further, so that it wasn't simply the fact that you partook in such behaviors—or even that you did so frequently—but that your urges were interfering with other important (nonsexual) activities and obligations, such as, you know, eating meals, bathing, remembering to pick up the kids from school, that sort of thing. Kafka knew from many years of experience with treating so-called sex addicts that patients' unsuccessful efforts to curb their problematic habits could cause tremendous difficulties in their lives. And it was this distress over what the *patient* perceived to be his or her own erotic overindulgences that motivated Kafka to push for "hypersexual disorder" to be added to the *DSM-5*. In fact, I think his position on these grounds—that is to say, basing the diagnosis on the subjective negative experience of the patient, not on some attempt by psychiatrists to ob-

---

*Note that just a few short decades ago, "culturally tolerated . . . homosexual behaviors" would definitely not have been part of that sentence.
†People with deviant sexualities (such as zoophiles, pedophiles, fetishists, exhibitionists, and voyeurs) can also be at the "hypersexual" end of the spectrum (in fact, individuals with such dispositions typically have unusually high sex drives), but Kafka's proposed "hypersexual disorder" applied only to "normophiles."

jectively define sexual excess—is both reasonable and based on admirable intentions.

Unfortunately, it's also conceptually problematic. The trouble is that much of the patient's malcontent (or in clinical terms his "personal distress") comes from being a member of a sexual minority, in this case one that's characterized by having a much higher sex drive than the statistical average. There were once countless gay men and lesbians with similar feelings of personal distress about their uncontrollable desires, after all. They were even diagnosed for a while with an APA-certified mental disorder called "ego-dystonic homosexuality" (those who were gay, in other words, but really, *really* didn't want to be). In reality, it's more often the society in which gay people have lived that is disordered, not them. Kafka's "hypersexual disorder" invites similar concerns. The only difference is that hypersexuals are led to see themselves as sick not for *whom* or *what* they want to have sex with but for *how often* and *how much* they want it.

In understanding that the patient's inner reality (especially the experience of personal distress) is a reflection of the society in which he or she lives, you can see how, deep down, Kafka's criteria aren't entirely absent a moral bias regarding the sexually "appropriate." In the days when nymphomaniacs ran amok, doctors and the public alike were convinced of what was appropriate. Most of us are now willing to say that Victorian physicians were sorely mistaken, with doctors confusing morality with medicine, but there's certainly no guarantee that hundreds of years from now people won't similarly be looking back at our present consensus on the sexually "appropriate" while shaking their heads over our own backwardness. You can pretty much count on that, in fact.

Kafka's unsuccessful bid to add "hypersexual disorder" to the *DSM-5* was due in no small part to the history of nymphomania and satyriasis being mired in moralism. When it came time for committee members to evaluate his proposed addition, many of his colleagues had become unwilling to take a medical stand on the issue. The psychologist Charles Moser, for example, pointed out that those inclined to divide the "sane" from the "insane" in terms of frequency of sex and intensity of desires overlook the possibility that sex itself may be the most meaningful part of a person's life, "which *appropriately* can take precedence over other activities" (italics added). Casanova may not have been entirely the virtuous womanizer many like to think he was, for instance, but it was indeed the pursuit of sex above all else that enriched his life and inspired his legend. "Cultivating whatever gave pleasure to my senses was always the chief business of my life," he wrote in his memoirs. "I have never found any occupation more important."

One person's sexual exorbitance is another's slow Monday morning. In my opinion, the only important point to weigh when trying to decide what is or isn't sexually appropriate is, again, that of harm. Deviations from population-level averages are useful in helping us to understand the full range of erotic diversity, but as we've seen, the issue of what is "normal" or "natural" is as shallow as a thimble when it comes to guiding us about how to behave. Unless we want to invoke the idea of a God who wound up the clockwork settings of an evolved human nature (and there's certainly no reason to do *that*), normal is just a number. And it's one without any inherent moral value at all.

Normal is still an interesting number, however. So for the sake of comparing a "normal" sex drive with a "hypersexual" one, we should start with the *average* number of orgasms that the *average* human being has over the course of a given week. Let's have a look at men first. Some of the best estimates of male ejaculatory frequencies can still be found in Alfred Kinsey's *Sexual Behavior in the Human Male* from 1948. Kinsey's book, coauthored by Wardell Pomeroy and Clyde Martin, set the gold standard for the objective study of human sexuality. Among other things, their carefully collected data with thousands of randomly sampled participants showed that 75 percent of adult American males were technically "sex deviants" according to the mental health criteria at the time. One of the many topics explored by Kinsey and his associates involved something they called TSO (or "total sexual outlet"). TSO was a man's cumulative total number of orgasms achieved per week by any single sex act or by some combination of sex acts. Whether he relied on old-fashioned means for ejaculation like masturbation or marital sex, or included something a bit more risqué, like inserting his penis into one of those newfangled vacuum cleaners, was inconsequential to the calculation of a man's TSO. It was simply how many times a given man living in the United States in the year 1948 expelled seminal fluid over the course of an average week. In a sample of fifty-three hundred males of reproductive age, the average TSO was 2.5. (Not that such a thing doesn't occur, but note that's an aggregate mean— the ".5" doesn't represent a half orgasm.) But some men in the study were more Casanovian in their physiologies, reporting ejaculation frequencies that were well above this average. In fact, about 8 percent of those sampled had a mean TSO of 7 or more for five consecutive years. That translates to at least one orgasm a day, every day, for the past half decade.

Much has changed since Kinsey collected his data, such as relaxing attitudes about masturbation, not to mention the ability to summon up nude strangers on our laptops whenever we want to. "In my day" (as old-timers like me in their late thirties can now say), the best I could do as a horny teenager was to waste an entire afternoon glued to an exercise-equipment infomercial showing an overly tanned, equine-toothed spokesman doing sit-ups while wearing tight red shorts that promised a big reveal at any moment but, alas, never delivered. It's not that I didn't fantasize about more explicit scenes. Oh, the things I dreamed up. But like most emerging adults in the early 1990s, I wasn't about to go on a public scavenger hunt for hard-core porn. When I turned eighteen, I slid shamefully into an independent bookstore and blushingly purchased an expensive Robert Mapplethorpe volume (whose images of anal fisting and pierced frenulums were *slightly* more extreme than what I was used to), but that was about as far as I was willing to go to attach my face and reputation to my self-consciously sordid desires. Then the Internet happened, and, well, let's just say I haven't needed to watch any of those lame infomercials since.

The advent of the Internet created the perfect set of conditions for the consumption of porn, what scholars refer to as the "Triple-A Engine" effect: affordability, access, and anonymity. According to a set of statistics from 2010, 40 million Americans are regular consumers of porn. One in every three of these people is a woman. Thirty-five percent of all Internet downloads is porn related. Each day, 2.5 billion e-mails containing porn are sent and received, and 68 million search engine requests (25 percent of all terms entered) are for porn. Around 12 percent of all websites are pornographic.* Ninety percent of children aged 8 to 16 have viewed porn for "adults only" online, most while "doing their homework." The average age of their first exposure to this content is 11. Whether you're a fan of porn or see it as a plague, these are simply the naked facts.

Yet even in the face of tectonic shifts in culture and technology, Kinsey's old TSO figures have proven remarkably stable over the many decades since he first sat down to interview men about their hidden sex lives. (If anything, evidence shows that TSO counts have fallen a bit since then.) In 1987, for example, a study revealed that only 5 percent of high school boys and 3 per-

*Precise figures on Internet porn are notoriously contentious, largely varying by the methodology used to ascertain such data. In their book, *A Billion Wicked Thoughts: What the World's Largest Experiment Reveals About Human Desire* (New York: Dutton, 2011), for example, the neuroscientists Ogi Ogas and Sai Gaddam give a more conservative estimate of around 4 percent of all websites being pornographic.

cent of male college students masturbated on a daily basis. A 1994 survey of American men aged 18 to 59 reported that 1.2 percent routinely masturbated more than once a day. In 2002 (well into the Internet era), another study showed that the average male undergraduate masturbates about three times a week.* When it comes to sex that involves parties other than one's own palm, findings from the 1994 survey indicated 7.6 percent of American males were then engaging in partnered sex (married or unmarried, gay or straight) four or more times a week.

There are biological limits to the number of orgasms men can possibly have within a period of time, but even so, means and averages are linked to the cultural contexts in which they occur. Consider how Kinsey's concept of TSO would apply to the Aka people of central Africa, a peaceful group of foragers. In 2010, the anthropologists Barry and Bonnie Hewlett were startled to discover that married Aka couples have sex four or five times *in a single night*. It's not that they don't enjoy it also, but the Aka view sex primarily as a form of labor, quite literally a "search for children." "The work of the penis is the work to find a child," said one informant. "I am now doing it five times a night to search for a child," said another. "If I do not do it five times my wife will not be happy because she wants children quickly." Even after the woman is pregnant, this astounding hypersexuality continues due to the Aka's belief in "seminal nurture," which is the idea (similar to the semen-ingestion ritual of the Sambia in Papua New Guinea thought to transform boys into warriors) that semen is an essential "milk" or nutritive substance that all fetuses need in order to develop normally. Repeatedly inseminating their pregnant wives is a culturally mandated sex practice for Aka

*The accuracy of these figures can always be doubted, of course, since it's not as if there were hidden cameras in men's bedrooms. Still, ensuring participants' anonymity and confidentiality, combined with similar rates across multiple studies, tells us that these TSO figures are reasonable approximations of what's happening behind closed doors.

men.* They do take a few nights off during the week, and the husbands use a kind of natural Viagra in the form of a mysterious tree bark called *bolumba*, said to go down best with palm wine. Furthermore, once the baby is born, there's a moratorium on sex altogether for a while. Still, the TSO counts for married Aka males make the rest of us men look like monks.

It's unclear if the Aka wives' orgasms are matched in frequency with those of their husbands (they don't *need* to climax to conceive, after all), but we do have some sense of how often American women are having orgasms—or at least how often they were doing so in 1953. That's when Kinsey's *Sexual Behavior in the Human Female* came out.† Kinsey and his colleagues found that unlike for male ejaculation, the female TSO was a function of marital status. Unmarried women had a mean of 0.5 orgasms a week, whereas the average for married women was 2.2.

Aside from those old sexist books written by Ellis and others, there are still virtually no studies on female "sex addicts," a reflection of the fact that women represent only a small fraction of cases presenting to clinicians. (Kafka points out that there are five male hypersexuals for every one female, a statistic also in keeping with the parental investment theory basics we saw in the previous chapter.) In a 2006 survey of 1,171 Swedish women, 80 of them (around 7 percent) were labeled "hypersexual." Why the researchers settled on thirteen orgasms per month as the critical dividing line between "normal sexuality" and "hypersexuality" in women is something of a puzzle (there's nothing special or catastrophic about that figure so far as *I* can tell), but nonetheless any *kvinna* finding herself on the wrong side of that line was consid-

---

*The Aka's belief in seminal nurture can be contrasted with that of the Gikuyu people of Africa—and of many Westerners too, for that matter. The Gikuyu believe that penile-vaginal intercourse during pregnancy can harm the fetus (it can't, by the way), and so Gikuyu men with expectant wives are required to tie up their foreskins into a special tassel called a "bush," allowing no more than two inches of penetration.
†The "Human Male" and "Human Female" parts of Kinsey's titles should really be placed in scare quotes today, given what we've since learned about the range of cultural variation in the sexual practices of other human societies such as the Aka.

ered "hypersexual." The bar for the Swedish male respondents in the same survey was set somewhat higher. Men needed a minimum of seventeen orgasms a month (another dubious figure) to be classified by the researchers as "hypersexual." And 151 of the 1,244 men responding to the survey (around 12 percent) could indeed claim that title.

In addition to simply climaxing more frequently than their peers, these hypersexual Swedes—male and female, gay, bisexual, and straight—had other things in common with one another as well. Compared with the nonhypersexuals, for example, they reported becoming sexually active at an earlier age, expressed a wider range of sexual preferences, paid more often for sex, and were more likely to identify as exhibitionists, voyeurs, and sadomasochists (I'll forever dream of IKEAs stocked floor to ceiling with nipple clamps and bondage hoods). But it's not all fun and games for Swedish hypersexuals. The women in this category were more likely to report having been sexually abused, too. Overall, both the men and the women were also less satisfied with their sex lives than the nonhypersexuals and reported more relationship problems and STIs. They were also more likely to have sought professional help for sex-related issues.

But this is where that personal distress problem from before comes into the picture, because these negative effects were true only for those hypersexuals having a lot of "impersonal sex," a term used by the authors to describe sex with anyone other than a committed partner. By contrast, those hypersexuals having loads of "personal sex" were the happiest respondents of all.* What this tells us is that hypersexuality isn't the cause of a per-

---

*Religious upbringing and related beliefs about sex also play a moderating role. One study found that male hypersexuals who sought treatment were likely to be members of organized religions and to report that religion was important to them. See Rory C. Reid, Bruce N. Carpenter, and Thad Q. Lloyd, "Assessing Psychological Symptom Patterns of Patients Seeking Help for Hypersexual Behavior," *Sexual and Relationship Therapy* 24, no. 1 (2009): 47–63.

son's problems per se; rather, it's the way that people express their high sex drives that determines whether they'll run into difficulties. Like other sexual minorities, most hypersexuals find themselves in societies that fail to tolerate harmless, biologically based erotic differences. For those who aren't in relationships (or for those who are, just not with a fellow hypersexual), having the urge or need to orgasm several times a day can lend itself to pursuits of casual (or "impersonal") sex. Having frequent no-strings-attached sex (or being "promiscuous") is still frowned upon even in societies as relatively progressive as Sweden's. So for hypersexuals to report being "less satisfied" with their "impersonal" sex lives and having "relationship problems" and "sex-related issues" under such social conditions really isn't all that surprising.

❧

There *are* situations, however, where hypersexuality is a genuine medical crisis. In some cases, an unusually febrile and chronic state of desire can appear almost overnight in an individual who doesn't normally have a particularly high sex drive. When that happens, it can indeed be a worrisome symptom of a serious underlying health problem. After all, localized brain regions govern *all* thought and behavior, including those related to our erotic responses. Our personalities and our sex drives can be dramatically altered by physical injury to certain brain regions or by drugs that modify our neurochemicals, as well as by infections and viruses that seep into our cerebrums. So when an average Joe (or Jane) suddenly morphs into the Marquis de Sade, this could be a sign of serious engine troubles under the hood.

Consider the case of a twenty-eight-year-old housewife from India, for example, who noticed a sudden change in her own libido. Her sex drive was off the charts, she found herself persistently aroused, and she'd been having multiple orgasms for the past two days. Since it was so unlike her (and since this unyielding state of yearning had begun to cause the type of marital friction that she definitely *didn't* desire), she sought the help of her gynecologist. Her physician couldn't explain the woman's hypersexuality, either. But after a few days of scratching his head, the gynecologist's colleague, an epidemiologist from a nearby hospital, eventually solved the mystery. It turned out that a rabid puppy had bitten the woman a few months earlier while she played with the dog at a neighbor's house. She didn't know it was rabid, and so she'd brushed off the incident as a minor event, as most people would. (It was just a small puppy, after all, not exactly Cujo.) Unfortunately, by the time anyone managed to figure out that the abrupt spike in her passions was actually a warning sign of the rabies rampaging through her brain, it was too late. "She expired on the fourth day," the doctors close their sad case report.

Rabies is just one of many distinct neurological conditions for which hypersexuality may appear as a rather jarring symptom. It's not inevitable, and by and large most people with these conditions *don't* become hypersexual, but extreme libidos have also been found in patients with Tourette's syndrome, multiple sclerosis, Huntington's disease, and, most notably, Klüver-Bucy syndrome. Klüver-Bucy is a rare disorder that can be brought on by a variety of factors, including herpes encephalitis and oxygen deprivation. The condition has likely been around ever since there were brains to go haywire, but it was first discovered in 1939 when the epilepsy researchers Heinrich Klüver and Paul Bucy were tinkering inside the skulls of live rhesus monkeys to better understand seizures. When the scientists removed chunks of these poor animals' medial temporal lobes, where we now know the central problem behind this condition lies, the monkeys became so sexually agitated that they'd start gyrating against the operating table. Since then, Klüver-Bucy syndrome has been the subject of several high-profile legal cases in which a person with the condition commits a sex crime and the judge must then decide if the accused is to be held responsible (that is, punished) for his or her acts. This is exactly what that other

judge did so long ago with Krafft-Ebing's "Clemence" on his attempted rape charges. And it's disconcerting, actually, that even
with psychiatrists' expert testimonies about hypersexuality being a known indicator of a malfunctioning brain, most judges
presiding over such cases today have been considerably less lenient toward defendants with Klüver-Bucy syndrome than a
judge in 1886 was with a man suffering from "satyriasis."

Incidentally, Krafft-Ebing was probably on the right track
about our old pal Clemence in the late nineteenth century. When
the out-of-control middle-aged engineer who'd stepped off that
train in Brück and assaulted the elderly woman was five years old,
he'd been accidentally struck in the head by a hoe (of the farm
tool variety, just to avoid any confusion). In Krafft-Ebing's notes
from his physical exam, he mentions Clemence's right parietal
and frontal bones being still noticeably dented some forty years
later—"the overlying skin was united to the [skull]"—and when
the author of *Psychopathia Sexualis* applied pressure to this spot
on the patient's head, sparks of pain irradiated the lower branch
of Clemence's trigeminal nerve. Neuroscience was in its infancy
then, and so Krafft-Ebing didn't see anything especially relevant
about the head injury at the time. But it's revealing to us now,
because this cortical area has since been implicated in brain-
damaged patients' difficulties with behavioral inhibition.

Sudden changes in a person's normal sex drive can clearly signal
a major underlying health problem, but "conditions" such as
"madness from the womb," "nymphomania," "satyriasis," and,
more recently, "hypersexual disorder" have been conceptualized
as diseases in their own right. Other than reflecting our own
moral biases, the practice of medicalizing these erotic outliers
has arguably done much more harm to those diagnosed as such
than good. As we've seen, not everyone has equivalent TSOs,
and any such scatter plot of lust is simply a display of biological

variation for this particular human trait. The variance of our natural libidos (even the points at the extreme ends) is the same as differences in our skin color, the shapes of our noses, or our tolerance for dairy products. Just as it's illogical to judge a person for his or her unique expression of those traits, it makes little sense to ascribe *moral* significance to the way a person's genes happen to be expressed on this particular spectrum of sexual diversity.

Whatever your views on hypersexuality (or "excessive expressions of culturally tolerated sexual behavior"), I'm willing to bet that you noticed aspects of yourself in one or two of the "normophiles" appearing in this chapter. The likelihood of you empathizing with the characters we're about to meet in the chapters that follow may be considerably less. Look a bit closer, though, and look honestly, and I have a feeling you'll see glimmers of yourself in them as well.

FOUR

CUPID THE PSYCHOPATH

Winged Cupid, rash and hardy, who by his evil manners, contemning all public justice and law, armed with fire and arrows, running up and down in the nights from house to house, and corrupting lawful marriages of every person, doth nothing but evil.

—Lucius Apuleius, *The Tale of Cupid and Psyche*
(late second century A.D.)

I didn't fear Cupid when I was a child, but little did I know he was a malevolent demon. By the time I first came to know him in the 1970s, Cupid's dark side was unrecognizable, having long since been painted over with baby fat and giggles by Hallmark and other great cultural bowdlerizers. This dramatic transformation of evil Cupid into a jubilant cherub, pink as a piglet's bottom and sweet as a lamb, who spies you each approaching Valentine's Day with twinkling eyes from the greeting-card display of your local pharmacy, who twirls with his arrow drawn ever so gently in the form of a paper mobile pinned to the ceilings of third-grade classrooms, bears faint resemblance to his original manifestation. He still strikes me as a bit cheeky, but the real Cupid was a genuine psychopath.

It was the Roman writer Lucius Apuleius who brought Cupid to life in his ancient book of fables, *The Golden Ass*. Apuleius's Cupid was no mischievous toddler with hummingbird wings but an impulsive god who rejoiced in causing sexual havoc for all earthly creatures. Even the fearless Apollo refers to Cupid as "serpent dire and fierce":

> Who flies with wings above in starry skies,
> And doth subdue each thing with fiery flight.
> The Gods themselves and powers that seem so wise
> With mighty love be subject to his might.
> The rivers black and deadly floods of pain
> And darkness eke as thrall to him remain.

Apuleius's story reads like a second-century black comedy. Cupid's mother, Venus, is jealous of a mortal girl named Psyche whose beauty surpasses even her own. Seeking vengeance, Venus recruits her son, this bringer of misplaced desire, to shoot one of his fearsome poisoned arrows into the young maiden so that she'll forever covet the first reproachful thing she sets eyes upon, thereby punishing Psyche's "disobedient beauty" with a shameful attraction. "I pray thee without delay," says Venus pleadingly to her son Cupid, "that she may fall in love with the most miserable creature living, the most poor, the most crooked, and the most vile, that there may be none found in all the world of like wretchedness." Cupid breaks into Psyche's house one night and fully intends to do just this for his mother, but in the end he becomes so smitten with Psyche that he accidentally pricks himself with his own poisoned arrow while watching her sleep. This forever binds the two and leads to Psyche's entry by marriage into the Roman pantheon—and, to put it mildly, a rather tense relationship with her new mother-in-law.

Love may have saved him in the end, but Apuleius's Cupid wasn't so much a romantic matchmaker as a devil subjecting hapless people to a toxic lust, one that blinded them with the sorts of hypersexual urges we saw in the previous chapter. This allegory of a capricious god who pierces mortal hearts only to burden them with some scandalous attraction out of sheer boredom or as favors to other gods is reminiscent of nature's cold mindlessness when it comes to the paraphilias. Individuals with the most deviant desires have similarly found themselves at the whim of a terrible randomness.

During the 1920s, many also found themselves lying on a couch in the office of the German psychiatrist Wilhelm Stekel, who saw their existential plight as anything but unfortunate . . . for him. The most exotic "perverts" always put a smile on his face. "Variatio delectat!" Stekel marveled in his nightmarishly titled book, *Sexual Aberrations*. "How innumerable are the variations which Eros [Cupid's original Greek name] creates in order to make the monotonous simplicity of the natural sex organ interesting to the sexologist." It was Stekel who coined the term "paraphilia." The first part of the word, *para-*, is Greek for "other" or "outside of," and *-philia* translates roughly to "loving."*

The son of an illiterate Orthodox Jew and an unpleasant but at least educated mother, Stekel had trained briefly under Krafft-Ebing and was a former friend and apostle of Sigmund Freud's. The two had a massive falling-out when the indiscreet and famously ornery Sigmund spilled the beans about Wilhelm's own mysterious perversion to a mutual colleague, the psychoanalyst Ernest Jones. It remains mysterious to our inquiring minds

---

*Some sexologists have since pointed out that "paraphilia" is in fact a misnomer when applied to deviant desires. They prefer to use the more arcane term "paralagnia," since *lagnia* is Greek for "lust." This is indeed a better fit given that the main issue is what stirs people's genitals, not love per se. Both love and lust are basic ingredients that must be mixed together in order for any decent romance to occur, and they can affect each other in fascinating ways. But neither is dependent on the other, and it's only lust that we're focusing on here. In any event, with these minor points cleared up, we'll be sticking with the more common "paraphilia" and its typical use in describing unusual sexual desires.

because Ernest proved to be a much more reliable confidant than Sigmund. Whatever it was, Stekel's erotic peculiarity went with him to his grave in 1940, an event hastened through suicide when he overdosed on aspirin in a London hotel room at the age of seventy-two to avoid a gangrenous foot having to be amputated due to his diabetes.*

This pitiable ending caught my attention for two reasons. First, I'm diabetic as well, and in my neurotic imaginings of a distant future I've been dispensed to some godforsaken outpost of geriatric infirmaries, and there I sit alone, day after day, in a room made bleary by retinopathy while rubbing the smooth stumps of my amputated legs. So I can commiserate with Stekel as he pondered a similar fate in that lonely hotel room in London. The second reason that this depressing denouement to Stekel's life story registered in my mind is that eighteen years before he swallowed that fatal dose of aspirin to escape the prospect of an amputation, he'd ironically chosen to highlight the case of an amputee fetishist in *Sexual Aberrations*. The fetishist was an eloquent physician whose dreams involved bedding the ultimate prize: a "pretty young girl who has been amputated at the thigh and wears a wooden leg." Like many with paraphilias, this fetishistic doctor was quite picky about the specific details, too. "The sight, image or even intercourse with a woman amputated on both sides," he clarified, "would leave me absolutely cold."

Such lovers of lost limbs are now politely called "acrotomophiles" (from the Greek, *akron*, extremity + *tomein*, to cut), but amputee fetishists have been around for some time. We know

---

*At seventeen, I distinctly recall promising to myself how I'd "go to my grave" without telling another living soul that I was gay. (Except, perhaps, for the closeted lesbian that I'd marry and with whom I'd swear a blood oath to never, ever, under any circumstances, reveal our dark secret. Preferably, she'd be on the tomboyish side just to make it doable.) Two things changed me. First, I almost died myself in an Atlanta hotel room of an accidental insulin overdose, and this spurred me to live a more open life. Second, science: the power of reason over shame and fear gave me a good hard shove out of the closet.

they date back to at least 1890, which is when a Chicago surgeon named G. Frank Lydston described a patient obsessed with a woman whose right leg had been removed. They'd dated for a while, and once they split up, the man was only interested in seeing other young ladies with a similar defect. In his original write-up on the subject, Stekel also describes others with this strong attraction to amputees, and he refers to it as the "negative fetish" (not in a judgmental way, but simply as a matter of bodily arithmetic). It's tempting for us to assume that deviant desires such as those of the acrotomophiles are merely the products of today's online carnivalesque porn (and there are indeed plenty of adult websites catering to acrotomophilic lust, which some aficionados playfully refer to as "monopede mania").* But given the preexisting history of *most* of the paraphilias, it's more likely that the Internet has served to pull together marginalized people who'd have otherwise been sexually isolated rather than to create fundamentally new kinds of paraphilias altogether.† (Another

---

*A close cousin to the acrotomophile is the apotemnophile, whose erotic fantasies involve having his *own* arm or leg removed. In 2005, the psychiatrist Michael First interviewed fifty-two middle-aged subjects (all males) who reported a lifelong desire to have a limb amputated. Fourteen had already had one removed. Nearly all of the subjects were gainfully employed and college educated. Much as transsexuals say they feel as if they were born into the wrong body, apotemnophiles just can't shake the feeling that their "true self" is an amputee. (The most envied alteration by far was an above-the-knee leg job.) Many were so desperate that they'd gone to extraordinary lengths to rid themselves of the misfit body part. The psychiatrist uncovered people who'd crushed their legs under gym weights or used shotguns, chain saws, a wood chipper, and even dry ice to liberate a rudely extraneous appendage. If they caused enough damage, many reasoned, a surgeon would be forced to finish what they started. The biggest regret for those who'd managed to escape their burdensome appendage was that they hadn't done it sooner. "I am absolutely ecstatic; I'm in possession of myself and my sexuality," said one. "It finally put me at peace," said another. "I no longer have that constant, gnawing frustration." See Michael B. First, "Desire for Amputation of a Limb: Paraphilia, Psychosis, or a New Type of Identity Disorder," *Psychological Medicine* 35, no. 6 (2005): 926.

†Yet paraphilias do change with the times. A once relatively common type of paraphilia known as agalmatophilia (from the Greek *agalma*, statue) has all but gone extinct today. Frequent references to some men's exclusive sexual interest in stone statues can be found throughout antiquity, especially in the records of ancient Rome and Greece. Pliny the Elder wrote of a man who fell in love with a statue of the goddess Aphrodite

reason to doubt the latter hypothesis, as we'll see, is that the psychological origins of paraphilias are usually traced back to a person's early childhood, long before these people go online for porn.)

Just knowing that there are acrotomophiles out there brings me comfort, though. Whenever I hear woeful tales like that of Wilhelm Stekel's, I have only to remind myself that with every toe that blackens from the syrup gurgling in my diabetic veins, I'm being slowly made over into a ravishing beauty for someone out there with just the right configuration of urges. To that gay acrotomophile, my losing a foot would be no less than an aphrodisiac, just as much so as was having it attached for my old podophilic friend. (He surprised me before, so it's hard to say, but I suspect he might have drawn the line at sucking on *gangrenous* toes.) The thought that I'd lose a body part but gain an admirer isn't *that* consoling, now that I think about it. Nonetheless, it's entirely true that there are people in this world for whom the seemingly rigid standards of beauty—youth, a nice complexion, being in good shape, a full complement of limbs— just don't apply.

It's been well over a century since they were first examined, but scientists still know astonishingly little about the paraphilias,

---

and "hiding by night embraced it [so] that a stain betrays his lustful act." See A. Scobie and A. J. W. Taylor, "Perversions Ancient and Modern: I. Agalmatophilia, the Statue Syndrome," *Journal of the History of the Behavioral Sciences* 11, no. 1 (1975): 51. The agalmatophiles' modern descendants lust after realistic life-size dolls (pediophilia, from the Greek *pedio*, doll; not to be confused with pedophilia). A virtual explosion in the ranks of the robotophiles is right around the corner. We may have lost agalmatophilia, but advances in technology mean that we've since gained everything from latex fetishism to mechanophilic arousal by automobiles to the electrophile's sexual dependence on electric currents. It's not just an ever-accelerating technology that broadens the paraphilic range, but changes to social conditions can modify, distort, or alter the human form in ways that similarly introduce new possibilities for sexual imprinting. During times of war, for example, amputees are more common a sight than during extended times of peace.

whether they involve an attraction to amputees, horses, corpses, feet, or whatever else.* We can certainly describe the paraphilias in graphic, shocking detail (which makes us no more than gawkers at a sideshow), group them into exotic genuses and families (which is what many psychiatrists specialize in doing and will soon come in handy for getting our ducks in a row), and maybe even trace the trajectory of a paraphilia from the person's childhood (or its "etiology," in clinical terms, which psychoanalysts sometimes attempt to unravel). But with all of these approaches, many questions will linger, such as why one person but not another with a very similar set of genes and experiences becomes paraphilic, why paraphilias are so overwhelmingly a male phenomenon (with experts reporting a sex-difference ratio of as much as ninety-nine paraphilic men to every paraphilic woman), and why paraphilias are so irreversible once they're set in motion.†

These questions remain unanswered not for lack of interest or talent. To the contrary, the area has been drawing some of the brightest minds since the earliest days of sexology. Rather, the difficulty in solving these riddles lies in the nature of the questions themselves, one that makes them impossible to address through controlled experiments. I mean, think about it. Let's say you're a researcher hoping to crack the case of what makes one child grow up to become—oh, I don't know—a melissaphile (which sounds like a cute moniker for someone who only wants to be with women named Melissa, but in fact it's an erotic attraction to bees, with *melissa* being Greek for "honeybee") while

---

*At last count there were more than five hundred identified paraphilias. See Anil Aggrawal, *Forensic and Medico-legal Aspects of Sexual Crimes and Unusual Sexual Practices* (Boca Raton, Fla.: CRC Press, 2009).

†The one exception to this drought of scientific knowledge is with our present understanding of pedophilia, a subject that has, for obvious and important social reasons, dominated empirical research in recent decades and has led to something of a renaissance of data-driven studies. Given the importance of that topic, pedophilia and some of the other "erotic age orientations" will receive their own extended treatment in chapter 6.

another turns into an adult who *insists* there be no stinging insects around during sex.

Giving them that clichéd old story about the birds and the bees at a fortuitous moment might do the trick, but you can't exactly randomly assign one group of children to a "bee condition" (the experimental group) and another to the "bee-less condition" (the control group) and twiddle your thumbs while waiting to see if any of them ends up a melissaphile fifteen years later. First of all, it's unclear what the experimental manipulation should even, ahem, *be*. Perhaps it would involve parents switching on a continuous audio recording of bumblebees buzzing before tucking their little ones into bed each night so that the ambient sound coincides with their first erotic dreams? Or maybe a special honeycomb hive built into a corner ceiling of their bedrooms so that when they're innocently playing doctor with another child, the presence of the bees comes to be automatically associated with their lifelong sexual arousal. So you see the problem (I hope). The cause of melissaphilia, and of every other paraphilia, indeed exists, but how to establish that cause *ethically* is the bane of every science-minded sexologist.

Given that experiments capable of tapping directly into the root causes of the paraphilias are methodologically verboten, research into the childhood origins of sexual deviance has necessarily been limited to conducting interviews, doing detailed case studies, and administering questionnaires to people who are already openly paraphilic. These are all examples of qualitative studies (rather than quantitative studies, in which experiments generate data that can be converted to objectively crunchable numbers, whereas interviews, case studies, and questionnaires usually produce narrative responses for the researchers' subjective analysis), and they're limited in their ability to get to the bottom of things. It may seem straightforward enough to just ask adults about their sexual experiences, but unfortunately such methods are notoriously susceptible to human error and bias.

On the subject's end, everything from false memories, to bending the truth by telling the researchers whatever it is he or she thinks they want to hear, to issues of outright deception can hugely distort the real picture. Investigators, meanwhile, who ask leading questions, who word their survey choices ambiguously, or who interpret the subject's vague responses in ways that support their preferred hypothesis also make data useless. Good qualitative scholars anticipate these factors and stem them off with research designs that can minimize such flaws, but even the best qualitative study can't definitively determine causation. On the positive side, qualitative research provides far richer descriptions and more detailed accounts of the subject's own experiences and interpretations, giving us glimpses into a person's private mental life that would be impossible to see with behavioral experiments alone. Such qualitative approaches have produced invaluable information about the paraphilias and have led to plausible, convincing theories (some of which we'll be examining). So I suppose a reasonable caveat emptor is this: whenever a study depends solely on what people *say* about their erotic desires or their sex lives, rather than on what they actually *think* or *do*, buyer beware.

Now, before we move on to what scientists at least strongly suspect about the paraphilias, there's one important side note to this issue of using controlled experiments in studying the real-time germination of sexual deviancy. For better or worse, when it comes to creating deviant rodents or farm animals, the ethical barriers are considerably less imposing than they are for turning kids into melissaphiles. And indeed, some researchers have managed to plant and cultivate full-blown "paraphilia-like" traits in members of other species by manipulating specific aspects of their early development.

In one study, for instance, newborn male rat pups that were randomly assigned to nurse from a dam (or mother rat) with teats that had been artificially coated with a lemony scent grew up to be "lemonophilic" adults. They could only get erections and ejaculate with females that smelled like their citrusy mothers. In another experiment, scientists switched newborn sheep and goats at birth: The baby goats were raised by sheep, and the baby sheep were raised by goats. Once these animals were reproductively mature, they were reunited with members of their "birth species." Such an unusual development had a dramatically different effect on the males and the females. For the now-adult males, neither the sheep nor the goats displayed even the slightest interest in the female members of their own biological kind. Rather, their mating efforts were directed to females of their adoptive species. By contrast, the adult females who'd been switched at birth were far less discriminating in their choice of a male sex partner. Unlike the males', these females' potential for arousal went both ways when it came to their sheep-goat options. As long as it was one of the two, it had a penis, and it knew what to do with it, the particular species was just a superficial detail to them.

In other words, the males responded to their unusual development in a fundamentally different way from how the females responded to the *same* aberrant experience. The former underwent a "sexual imprinting" process that wired their arousal pattern in one direction only (toward those they were exposed to while growing up), whereas the latter retained significantly more erotic plasticity. Rats, sheep, and goats are many branches away from us on the phylogenetic tree, so there's only so much we can make of these findings when it comes to paraphilias in our own species. But keep these animal studies in mind (better yet, earmark this section so you can come back to it), because we'll soon be encountering examples of what certainly *look* to be sexual imprinting in early human development, and there are also a few differences between men and women that are eerily consistent with the ungulate gender-difference pattern we've just seen.

Rest assured, there are no warped *Jungle Book* tales lying in wait for you. As far as I'm aware, for instance, no little girls have ever been raised by a community of bonobos, *let alone* any who eventually grew up to grind their clitorises with female apes. Yet I'd be remiss not to briefly share with you the story of Lucy, a chimp raised much like a human child by the psychologist Maurice Temerlin in the late 1960s. Temerlin reports that once she'd blossomed into a young adult, a favorite hobby of Lucy's was masturbating to the nude *human* male centerfolds from the latest issue of *Playgirl* magazine by carefully spreading out the pages on the floor and then placing her swollen genitals on the image of the man's penis. (I'm a bit jealous, to be honest. My mom would only pick up the occasional *Men's Fitness* for me, and here Lucy was getting Temerlin to purchase a subscription to *Playgirl*.)

We human beings can't sacrifice a few of our own to be raised by another species just to see if it screws up a person's sexuality. Most parents, loving and responsible bastards that they are, aren't willing to donate their babies to such an ambitious project.* But we *can* be grouped in accordance with whatever makes us the horniest (or in other words what we're "into"). Wherever direct causation fails, taxonomy reigns. So before diving headfirst into the elaborate theories and conceptual issues involving the paraphilias, we should get squared away on some basic organizational matters.

Experts have sought to discipline this unruly subfield of psychiatry by classifying, carving up, and cajoling sex deviants into their own neat and tidy corners of the various diagnostic manuals. There are other templates for arranging the paraphilias, and sometimes also major differences between them (such as the

*Not that Lucy's parents were either, by the way. She was born wild in Africa. Poachers killed her mother when she was an infant, and Temerlin had adopted her as an orphan.

inclusion of "excessive sexual desires" in the *ICD-10* but not in the *DSM-5*). For the most part, however, they're more in agreement with one another than they are in disagreement. For simplicity's sake, we'll use the current scheme in the North American diagnostic manual (the *DSM-5*) to get a handle on what would otherwise be an impenetrable, unwieldy mass of outrageous case studies and forensic reports. If nothing else, this will provide us with a nomenclature of sexual deviancy, and when you're dealing with such a thorny subject matter as this, speaking a common language is essential.

The *DSM-5* includes only eight specific paraphilias: exhibitionism, fetishism, frotteurism (rubbing one's genitals or touching strangers in public, typically in crowded places such as subway trains or elevators), pedophilia, masochism, sadism, voyeurism, and transvestic fetishism. Rattling that list off out loud makes it sound—and unfortunately so—like a roundup for a "Things Your Mother Would Disown You For" survey on *Family Feud*. As we've already seen, there are many other varieties of paraphilias. They can be anything under the sun, "including the sun," as a psychiatrist once quipped.* But the core list is limited to these eight mostly for pragmatic reasons. If every florid fetish or atypical orientation were included in the *DSM-5*, there wouldn't be any room left for the many psychiatric disorders that *don't* involve the genitals.

Alas, there's a place in the *DSM-5* for even the most unusual paraphilias: the residual ninth category under the heading of PNOS, or "paraphilia not otherwise specified." Just as we saw with "hypersexuality" or "sex addiction" sometimes being diagnosed with the broad swipe of the *DSM-5*'s "sexual disorder not otherwise specified," most of the paraphilias aren't found explicitly in the *DSM-5*, either, but instead are tossed into this catchall PNOS

*And it's true, by the way. The term "actirasty" applies to those who become sexually aroused by the sun's rays.

bucket. A few of the "philias" that we've run into so far (such as zoophilia, necrophilia, podophilia, and acrotomophilia) could be diagnosed under this PNOS heading, and we'll turn our attention to several additional examples of these more deviant deviances in the pages ahead.

The taxonomy of deviance can get rather complicated. Consider that the same person may have multiple paraphilias, and these can overlap the seemingly snug boundaries laid out in the *DSM-5* in the most curious ways. For example, my first exposure to an erotic outlier was with a hebephilic pygophile who also had a touch of frotteurism. A hebephile (from "Hebe," the Greek goddess of youth) prefers pubescents between the ages of roughly eleven and fourteen, and as for pygophilia, it's all about the buttocks. Mix these two together and you get the character I happened to chance upon that afternoon long ago. It's a blurry reminiscence involving a rabbit-like rodent, a department store, and a man aroused by the scraggy body of the twelve-year-old boy that I inhabited at that time. My older sister had wandered off to another part of the store, leaving me alone in the pet section to pick out a calico-colored guinea pig that I'd already decided to call Trixie (which sounds to me now more like a streetwalker's nom de plume). I believe it was somewhere around the *fifth* time that the man asked me to bend over to take a closer look at the animals in their ground-level cages while—"Oops, *sorry*," he kept saying—casually running his hand against my backside that I got the vague sense that this whole business was about something other than our shared fascination with the species *Cavia porcellus*, that brave little emblem of so many bad medical metaphors. Given that he used his hands rather than pressed his penis against me—he did have the decency not to do *that* at least—a pedantic specialist might prefer the term "toucheurism" to "frotteurism." Both refer to unwelcome, surreptitious contact with strangers in public places for sexual gratification but differ by the appendage being used.

We shouldn't fault this man for his singular erotic profile any more than we should hold him accountable for the particular whorl of his hair pattern, but he hadn't kept his deviant desires in his head, where they belonged in this instance. Instead, they reached out and grabbed my ass. Now, the field of psychiatry has learned a few good lessons from its own awkward adolescent history with sex. For example, you don't call a woman a "nympho-maniac" just because she masturbates and likes to have sex, as Horatio Storer did, and you really shouldn't try to turn a gay man straight by encouraging one of your female patients to fondle him, as Albert Ellis did. This learning curve has led to the APA's rather lenient modern stance on the paraphilias, too. The basic rule is this: no matter how deviant your deviant desires, if you're not hurting anyone (or anything), and if your sexuality isn't causing you personal distress, then you don't meet the criteria for a men-tal illness connected to your paraphilia.

I wouldn't go so far as to say I was harmed that day. However, had my sister not reappeared right around this man's sixth grop-ing attempt, and given that I was so naive I might well have done whatever he told me to do (including follow him to his car, all seventy-five pounds of me), he was walking a precarious line even for my relatively liberal measures of harm. So regardless of whether he was perfectly comfortable with being a hebephilic pygophile whose favorite pastime is touching young boys in public places, and he experienced no personal distress due to his sexual nature, the average psychiatrist examining the store's security-camera footage from that day long ago would call this man men-tally ill. And the average citizen looking over that psychiatrist's shoulder, of course, would call him a criminal.

Of the eight specifically named paraphilias in the manual, five are set aside as exempt from the "personal distress" policy: pedophilia, exhibitionism, voyeurism, frotteurism, and sadism. The guiding thought is that whereas most forms of sexual devi-ance are harmless enough, these are inherently harmful. Yet the dividing line between "harmful" and "not harmful" isn't always so obvious. We'll explore this critically important issue in more detail in the next chapter. (And bear in mind that the question of whether someone has a mental illness is altogether different from that of whether he's broken the law. The issue is his ability to have behaved otherwise.*)

Leaving the matter of harm aside for the time being, let's return to taxonomy. The psychologist James Cantor suggests that we think of paraphilias as being divisible into two broad categories. First, he says, are those in which the sexually interesting object (what's commonly referred to as the "erotic target") is something other than reproductively mature, standard-issue human beings. The decadent French poet Charles Baudelaire claimed that he'd acquired a keen taste for female giantesses and dwarves, for instance. (The technical term for an attraction to people of dramatically different heights is "anasteemaphilia.") Then again, Baudelaire also once claimed to have eaten the brains of a child and that he owned a pair of riding breeches made from his father's hide, so we should probably take any claims of his love life with a good pinch of salt, too. But other flavorful examples fitting into this first category of unusual erotic targets would include

*One paraphilia for which this distinction between mental illness and criminal accountability has generated heated discussion is that of biastophilia. This is a subtype of sexual sadism where the erotic focus is centered on the coercive act itself, whether it's a particularly brutal act of rape or one that occurs by blackmail or threat. Arousal is triggered not by overt signs of injury, suffering, or humiliation per se but more narrowly by forcing intercourse upon an unwilling other. Only a handful of convicted rapists are believed to be "true biastophiles" (particularly repeat offenders and those with a history of violent crimes). In controlled studies, these men become more sexually aroused by video scenes and audio clips depicting rape than they do by those involving consensual sex. The exact opposite pattern is seen in the vast majority of men, *including* most rapists. See David Thornton, "Evidence Regarding the Need for a Diagnostic Category for a Coercive Paraphilia," *Archives of Sexual Behavior* 39, no. 2 (2010): 411–18.

"ornithophilia" (an intense desire for birds), "savantophilia" (arousal to mentally challenged individuals), and "chasmophilia" (the attraction to nooks, crannies, crevices, and chasms, more precisely, those that are *not* found on the surface of the human body).

The second broad category of paraphilias, Cantor goes on to explain, consists of those in which the most arousing sex act is something other than copulatory or precopulatory (or foreplay) behaviors with a consenting partner. Here you'll find the "stygiophiles" (who work themselves up into a hot lather at the thought of going to hell), the "psychrophiles" (who wouldn't mind going to hell either, but only once it freezes over, since their biggest turn-on is being cold and watching others shivering), and the "climacophiles" (who are said to have their most intense orgasms while falling down stairs).

These are all extremely rare. More familiar examples of this second category are frotteurism, exhibitionism, and voyeurism. The influential sexologist Kurt Freund, considered by many to be the father of experimental sexology (not to worry, we'll spend quality time with him later), saw each of these paraphilias as a sign of an underlying "courtship disorder." Freund argued that most healthy adult sexual encounters include a stage of "preparatory" behaviors that serve to convey the person's erotic interest. This isn't so much the foreplay stage as the one where green-lighting efforts are taking place. "This is the phase of eye-talk and finger-talk when the partners give signals or invitations to one another," the psychologist John Money clarified. "They flirt, coquet, woo, or lure one another. It is sometimes known as the phase of courtship or, in animals, the mating dance of display." Freund believed that some individuals essentially get "stuck" on a particular element of this courtship stage, and as a consequence they achieve their greatest sexual satisfaction at, say, the pre-intercourse phase of touching (frotteurism), or that of broadcasting their sexual interest in potential partners by showing their arousal (exhibitionism), or by acquiring privileged visual sexual access to another (voyeurism). According to the model, these paraphilias are exaggerated displays of ritualized courtship. As Money explains, each has somehow come to "push its way onto center field, instead of remaining on the sidelines. It displaces the main event, which is genital intercourse, and steals the spotlight."

Looking at Cantor's more general bipartite model of the paraphilias, however, we can see that the first category of paraphilias involves an unusual *subject* of desire, whereas the second centers on an unusual *activity*. It's a useful distinction in thinking about abnormal sexuality, but note that these categories aren't mutually exclusive. That is to say, someone could be paraphilic in both his erotic target *and* his favorite sex act. I mean, really, any psellismophilic nebulophile (someone whose most passionate moments involve masturbating in the foggy mist while listening to a person stutter) can see that.

Sexual fetishism is also a case where the division between erotic targets and erotic paraphilic activities can get somewhat blurred. As we're about to see, the fetish object isn't exactly the erotic target; instead, it's more of a symbolic stand-in for the real erotic target's genitals. Sexual gratification with the fetish object may or may not depend on a particular ritual, so it's also not a perfect fit for the category of activity-based paraphilias. Now, fetish objects can be anything imaginable. There have been cases of people fetishizing wheelchairs and crutches, hearing aids, rubber swim caps, and anything else that might serve as a sexual surrogate for the person's ideal partner. Some fetishists, such as the archetypal "panty bandit," are prone to theft of such objects, a secondary kleptomania that poses additional problems. Also, many paraphiliacs are unhappy to part with their special collections of fetish objects gathered over the years, guarding them like treasures. Stekel referred to such erotic compilations as the "fetishist

Bibles." An elderly man with a pubic hair fetish ("pubephilia," aptly enough) had an array so ancient that many of the wiry strands glued to the pages in his scrapbook had long since turned gray. (We can only speculate if those earlier principles of disgust management would apply to any unexpected encounters with the fossilized lice eggs hidden in that volume.)

For the fetishist, it's the object's physical connection with the erotic target's body, as though it has absorbed the person's hidden "essence," that makes it so arousing. This is an important point. Brand-new, never-before-worn pairs of shoes, for example, aren't likely to turn on a shoe fetishist; rather, he's only going to be interested in those shoes that have actually been worn by someone he finds appealing (or someone he can fantasize having worn them—a secondhand store is the fetishist's brothel). Likewise, if your fetish object is men's underwear, you're not going to tear into the plastic wrapping of that crisp new pair of Fruit of the Looms you picked up at Target the next time you're in the mood; you're going to pull out the semen-stained briefs worn by your favorite porn star that you won in that online auction the other week.

Even for nonparaphilic fetishists, it's easy to grasp the appeal of feeling or touching something intimately associated with a person you find attractive. I don't know about you, but *I'd* certainly be lying if I said that I've never been intensely aroused by a Diet Coke can. Back in high school, I once pleasured myself to the empty one discarded by a boy (a straight boy, alas) I had a crush on, since that was about as close as I could get to his lips: something of his essence, to my mind, must have magically been soldered onto it. And in consumer psychology, several studies have found that customers are more likely to purchase clothes that have been physically handled by good-looking salesclerks.*

---

*The effect can also be seen in a person's unwillingness to touch objects belonging to an undesirable person. For example, consider your comfort level in slipping on Jerry Sandusky's favorite sweater or wearing Jimmy Savile's glasses. Whether or not this disgust reaction is logical (it isn't, of course, since a person's spoiled reputation isn't

Casual sexual fetishism like this is quite common. What makes someone a "true fetishist" in the clinical, paraphilic sense is that the fetish object has become even more arousing than the actual erotic target that it represents and is now more or less *required* for sexual gratification.

The "objectophiles," also known as the "objectum sexuals," are similar to paraphilic fetishists in some ways, but in other ways they're altogether different. For these unique individuals, the lust-worthy object isn't a symbolic stand-in for the real erotic target, nor is their attraction to this item related in any way to its physical contact with a desirable person. Rather, for the objectum sexuals, the object *is* the erotic target. More important, they're absolutely convinced that the object has reciprocal sexual feelings for them. Chairs, ladders, shawls, bookshelves . . . you name it. To the objectophiles, any object in the galaxy has the capacity to fall madly, deeply in love (and lust) with particular human beings. (Worth bearing in mind, I suppose, the next time you go to kick your car's tires after it breaks down on you. If she's a masochist this might just encourage her.)

The psychotherapist Amy Marsh has suggested that objectum sexuals have a rare underlying neurological condition called "object personification synesthesia," which causes them to perceive personalities and emotions, including sexual desires, in inanimate objects.* These individuals are aroused by *individual*

---

literally a contagious pathogen that we can "catch"), it steers us away from "social contaminants" that are clearly signaling a perceived "health risk" to our own reputations. Expressions of symbolic disgust toward such individuals, especially when we're in the presence of others (wearing Sandusky's sweater or Savile's glasses in public is probably even less appealing to you than briefly donning them in the privacy of a lab), broadcast that you're "not like" the offending person. In the case of these two, for instance, their "child molester" essence is *the* most toxic symbolic substance in our society. See Paul Rozin and Edward B. Royzman, "Negativity Bias, Negativity Dominance, and Contagion," *Personality and Social Psychology Review* 5, no. 4 (2001): 296–320.

*They also commonly have Asperger's syndrome or fall somewhere along the autistic spectrum, neurocognitive conditions associated with impaired social functioning.

objects, not an entire class of fetish objects. In 1979, an objecto-
philic woman from Sweden named Eija-Riitta Eklöf made head-
lines when she married the Berlin Wall. Today, she considers
herself a widow. More recently, an American named Erika Eiffel
(you'll see why in a moment) was filmed for a documentary con-
summating her marriage to the Eiffel Tower. It's hard for newly-
weds to be intimate with each other when tourists are constantly
snapping photographs and ambling all around them, but there
she was, looking up devotedly at her loving 1,063-foot-tall
spouse. (She also sees the structure as a female, so it's a lesbian
relationship in that sense. I suppose one could say Erika is bisexual, given that her earlier relationship was with a male . . . a quiet
gentleman that most of us know as the Golden Gate Bridge.) In
the documentary, Erika lifted her trench coat demurely, strad-
dled one of the Eiffel Tower's massive steel foundations, and
sealed the coital deal. (There's more to Mrs. Eiffel than her objectum sexual identity, by the way. She's also an internationally
ranked professional archer.)

More commonly—if "commonly" can be used here—
objectum sexuals find themselves swooning over everyday objects, not flashy celebrities like the Berlin Wall and the Eiffel
Tower. In a 2010 study, Marsh interviewed dozens of objectophiles. "What does your beloved object or objects find most
attractive about you?" she asked them. One woman in a relationship with a flag named "Libby" replied: "Well, Libby is always telling me she thinks I am funny. We make each other
laugh so hard! . . . [It]'s hard to get a serious conversation out of
[flags], because they are always silly and joking around!" An objec-
tophilic man, meanwhile, mostly attracted to music soundboards,
remarked how his electronic lovers adored his physiognomy. "I'm
kind of a heavyset person, and they like that about me," he told
Marsh. "They like my hands. I have swan-neck deformity of my
fingers, they like that a lot. They also like the rough texture of
my fingers."

•

Speaking of hands and fingers, we shouldn't forget the partialists we met earlier, their ambassador being that podophile who deflowered my unsightly toes. Partialists aren't into objects so much as protuberances. You may recall that these are individuals for whom it's a specific part of the body—a part other than the reproductive organs—that captures their erotic fancy. One individual who most assuredly was *not* a partialist was Voltaire, that witty French philosopher of the Enlightenment, who made it perfectly clear that he was infatuated with every square inch of his mistress (who also happened to be his niece), Mme Denis, in a letter from 1748. "I press a thousand kisses on your round breasts," he wrote to her announcing his imminent arrival to visit her in Paris, "on your ravishing bottom, on all your person, which has so often given me erections." But Voltaire's "monotonous simplicity" in his cherishing the entire female form, Wilhelm Stekel would remind us with his wide-eyed wonderment over natural sexual diversity, is no reason to lose our joie de vivre.

After all, in *Sexual Aberrations*, Stekel introduces us to a twenty-eight-year-old German businessman known only by the abbreviation "P.L." For P.L., it was all about the hands. "Whenever a lovely female hand touches him, his erection is instantaneous," writes Stekel of this partialist. After some psychoanalytic sleuthing, he surmised that P.L.'s fixation could be traced back to the man's first sexual experience at the age of seven, which for uncertain reasons Stekel decides to inform us about in Latin: "He liked especially *cum nonullis commilitonibus mutuam masturbationem tractare.*" This just goes to show how profound a thing can sound when it's uttered in the language of jurisprudence, since any gravity disappears immediately upon translation: basically, P.L. enjoyed giving while getting hand jobs from other boys his age. Such mutual masturbation persisted into his late teenage years, in fact, with P.L. noticing that he was only attracted to other boys with pale, well-formed hands. Anything else left him limp.

Over time, Stekel explains, P.L. abandoned these gay days and (allegedly) became increasingly aroused by the thought of females, or at least the thought of feminine hands. ("No power in the world could make him touch another man's organ now," writes Stekel, "and even the sight of another's penis is disgusting to him.") We're not told, however, if he ever gave up fantasizing about the sight of a boy's hands touching *his* penis, which I'd bet my bottom dollar he didn't. In any event, aiming to settle down and start a family, P.L. met an enchanting young woman who he thought would be perfect for playing the role of his wife. She was ideal, "possess[ing] all of the feminine virtues," adds Stekel. But in a twist of fate worthy of Cupid's cunning, the girl bore exceedingly large hands and often had dirt caked beneath her nails, an obvious complication for any self-respecting hand fetishist. In the end, he just couldn't handle her hands. In my mind's eye, I see the image of a determined P.L. returning home one evening after an insightful session with Dr. Stekel: encircled by a ring of locomotive steam, he's on the train platform in the drizzling rain, placing one sure finger against the trembling lips of the beautiful woman who has come to greet him and saying, "It's not you, darling, it's me. Well, me and those damn meat hooks of yours."

Even the most comprehensive taxonomy of sexual deviance can't always foresee all the bizarre ways in which the paraphilias can possibly materialize, and sometimes it's unclear how a particular expression of sexual deviance should be labeled. Take the following case of an "oral partialism," for instance. Whether this was the most accurate diagnosis is debatable, I'd say. Yet oral partialism was how the Australian psychiatrists assigned to the patient chose to describe it. And as you're about to see, one can certainly forgive them any confusion over what, exactly, they should have called it.

Our hard-to-believe-it's-true-but-unfortunately-it-is story takes place this time in 1998 and involves a socially awkward, mor-bidly obese nineteen-year-old male. Clinically depressed and weighing in at more than three hundred pounds, the unkempt teen first came to the medical attention of doctors in a clinic in rural Australia due to his poor personal hygiene. The young man had developed ulcerated sores under his arms, above his pubis, and in his groin. His father, who should have dragged his son to the doctor's office long before he'd gotten to this point, finally brought him in and explained to the attending physicians how these purulent wounds had been festering for the better part of *five years.*

At first, rather surprisingly given the long duration of infec-tion, the prognosis was good. A standard course of antibiotics was prescribed, and assuming his hygiene improved, the doctors were optimistic that the adolescent's sores would fully heal. It turns out, however, that the patient wasn't especially keen to get rid of them. Odd as this may sound, he had fallen in love with these bubbling cankers. At a follow-up appointment, he showed poor compliance with taking his medicine, claiming that he'd lost the pills. When the hospital staff pressed him on the matter, he confessed to his sexual motivations for retaining his wounds. "The patient's primary fantasy stimulus was that of a woman's mouth," explain the psychiatrists eventually assigned to the case:

> The fantasy consisted of an image of the woman licking her fingers or gently biting her own lips. Behaviorally, the patient would simultaneously put his own fingers into the ulcers in the groin or under the arms and then lick the pus from [them] . . . He ingested the pus and found both the smell and taste exciting, although he was unable to pinpoint exactly the sexually stimulating aspect of this act.

This large, lonely soul demonstrates how the human imagina-tion can make even the most grotesque of elements subjectively

worthy of the most passionate longings. You've probably also noticed by now the distinctively male tenor of our examples. I do
wish I had more female paraphiliacs to tell you about, but remember that the most conservative estimate suggests a male-to-
female paraphilia ratio of 99 to 1. This enormous gender gap
appears across the almost infinite range of sexual deviance, too.
There are two exceptions. First are the objectum sexuals, given
that women appear to be just as likely as men to have the rare
object personification synesthesia that may be behind that paraphilia. The other is sadomasochism, where males outnumber females by only 20 to 1. (Yet within this latter subgroup, there are
far more female masochists than there are sadists. We'll discuss
S&M in painstaking detail in the following chapter.)

   This is also where those animal experiments come back into
play. Just like the sex differences in the cross-reared sheep and
goats, the robust sex ratio in the paraphilias betrays a similar difference in men's and women's sexuality. Human males whose
erotic brains are wired in their early childhoods to respond only
to *specific* cues in the environment resemble those male ungu-
lates that couldn't become aroused by their own biological kind,
but instead only by their adoptive species. Whether it involves
livestock or people, this sexual-imprinting process (in which a
highly circumscribed set of erotic targets is stamped early into
the individual's brain) appears to be a decidedly male characteristic. By contrast, the female sheep and goats that were able to "go
both ways" after their intensive cross-rearing experiences, equally
aroused by both their own biological kind and members of their
adoptive species, were exhibiting "erotic plasticity" (in which one
can be sexually excited by a wide range of stimuli). Interestingly
enough, erotic plasticity is also strikingly more apparent in human females than it is in human males. Another way to say this is
that a girl's developing sexuality is more fluid or labile (and for
once that's not a pun) than a boy's; it's less prone to getting locked
onto a specific category of erotic target during childhood.

The social psychologist Roy Baumeister once rounded up a half century's worth of data on the sexual differences between men and women, deriving from his analysis a sort of paraphilic axiom. "Once a man's sexual tastes emerge," he wrote, "they are less susceptible to change or adaptation than a woman's." The mountains of data he surveyed to arrive at that conclusion are indeed revealing of female sexual fluidity and the more confined range of desire in men. Self-described heterosexual women who are polyamorous, for instance, report that they almost always engage in cunnilingus (female oral sex) with the other women during group sex, whereas self-identified straight men from the same community almost never perform oral sex on each other. (In other words, while the "poly" in "polyamory" stands for "many" sexual partners, for male polyamorists, that *usually* doesn't mean having many sex partners with penises.) In fact, women in general are far more likely than men to report being bisexual. Tellingly, they're also more likely to change their self-identification as straight or gay during their adult lives. Furthermore, lesbians are more likely than gay men to say that their sexual orientation is a "choice," a term that really only makes sense, of course, if the individual is in fact bisexual and decides to commit to one label or another. (Incidentally, if an antigay bigot genuinely believes homosexuality is an intentional choice or a "lifestyle," then it stands to reason that person's frequent use of such words could very well be a linguistic reflection of his or her own bisexual desires.)

Work by the psychologist Meredith Chivers also illustrates a greater female erotic plasticity. In several studies, Chivers has found that both straight women and lesbians exhibit vaginal vasocongestion (or increased blood flow to the genitals, a response specific to female sexual arousal) to a surprising assortment of sexual stimuli. For example, a woman's genitals will respond this way not only to her *preferred* gender (which is to say, men for self-identified straight women and women for self-identified lesbians) but also to naked pictures of her *nonpreferred* gender.

They'll even become demonstrably aroused at this physiological level to video footage of other species having sex, notably graphic scenes of bonobo intercourse. That last finding has been replicated, so it wasn't just some quirky, happenstance overrepresentation of female zoophiles in the study. Chivers clarifies that women aren't always consciously aware of their arousal to such stimuli. At least they report not feeling as turned on as the objective state of their genitalia would otherwise suggest. The vagina has a mind of its own, in other words; I suppose anything's possible, but farther north in the female brain, a pair of frenetic bonobos getting it on probably doesn't top most women's list of sexy and hot.*

When Chivers ran the same kinds of studies with (non-paraphilic) men using the erection-measurement device of a "penile plethysmograph" (which we'll have a closer look at in chapter 6), she uncovered an entirely different pattern of genital responding. Basically, compared with women's, men's "southern brains" were more of the same mind as their "northern brains," in that their genitals lined up pretty much with how they'd describe their own sexual orientations. Penises of men who'd checked off the heterosexual box grew erect for salacious images of women and went limp at the sight of naked men, whereas those attached to the self-described gay men stood to attention for the photographs of nude men and withered in response to naked women. Yet even the raunchiest depictions of bonobo sex left the men of both persuasions entirely flaccid. In summary, these data told Chivers that unlike with women, there's not much of a rift between the subconscious and the conscious when it comes to men's sexual arousal.†

---

*There are no data on it at the moment, but it would be interesting to see how this split between a woman's subconscious and conscious arousal would apply to ovulating women, given those earlier disgust findings on "biologically suboptimal" sexual unions.
†Those men who attempt to hide their deviant desires for defensive social reasons, such as pedophiles desperately trying to conceal their attraction to children or homophobic

The evolutionary interpretation of these sex differences in arousal—and I'm aware of no other explanation that has been proposed—is that female genital hyper-responsiveness was biologically adaptive in the ancestral past. Back on the savannah tens of thousands of years ago, even the most unappealing sex cues would have often been followed by actual intercourse, so whether the woman wanted intercourse (it was consensual) or not (basically, rape), the capacity to become so easily physiologically aroused offered a sort of insurance policy against possible damage. Specifically, Chivers's "preparation hypothesis" posits that a woman's ready-for-anything genital arousal reduced physical injury to her reproductive organs by vaginal lubrication. "The costs of non-responding [genitally] to sexual cues, including nonpreferred cues," clarifies the psychologist Samantha Dawson, "would be expected to be much higher for women (e.g., tears and ecchymosis leading to infertility) than for men (e.g., the loss of a single reproductive opportunity)." In further support of this hypothesis, researchers have also found genital arousal to depictions of violent sexual coercion in women who consciously find the thought of rape revolting and terrifying, hardly erotic and arousing.

Needless to say, what we've reviewed so far are again relative findings concerning male and female sexuality, not inviolate statements of absolute differences between the sexes. There are exceptions to every generality, and Darwin got famous on this very premise of individual differences in biology. Still, as the work of Roy Baumeister, Meredith Chivers, and many others clearly suggests, if there *is* any degree of sexual imprinting in human females during development, for the vast majority of them

---

men trying to mask their own same-sex desires, have been found to exhibit the same circumscribed penile response pattern as those who are open about their sexuality. The only difference is that closeted men show the exact opposite pattern of arousal (the stigmatized) from whatever it is they claim to be attracted to (the socially acceptable).

it's much more easily overwritten than it is for human males.*
The paraphilic classifications in the *DSM-5* are gender neutral,
it's worth noting. So without a theory of innate sex differences to
account for the fact that this section of the diagnostic manual is
almost the exclusive province of male psychiatric patients, it's
rather hard to explain this overrepresentation on traditional feminist grounds of a sexist society. I've never heard complaints of
paraphilias being a male privilege and in fact I'm quite sure most
paraphilic men would be happy to share this wealth of shame
and stigma with the fairer sex.

Paraphilic males show a telltale range of erections just like
their nonparaphilic brethren; it's just that the focus of their attraction has somehow constricted around more unusual erotic
targets, which is something other than the norm. Many sexologists, and a lot of paraphiliacs themselves, believe that such
atypical arousal patterns link back to a specific event, or perhaps
a series of events, in the man's early boyhood. That's to say, sexual imprinting, just like what happened to those male rats that
suckled as pups from a set of lemony teats. The defining "imprint" in our own species seems to occur surprisingly early, usually being reported as sometime between the boy's fourth and
ninth birthdays, although it's best to think of this five-year time
frame as a "sensitive period" (with plus or minus several years on
either end) rather than as a "critical period" of male development. At puberty, the eroticized imprint is jogged awake by a

---

*Compared with plentiful retrospective accounts of paraphilic men, women's remem-
bered accounts of some definitive childhood experience they believe to be linked to
their adult sexual arousal are scarce. Yet a handful of such cases do exist. In 1960, for
instance, a woman recounted for a sex researcher what she considered to be the seeds
of her masochism: "When I was four, my father once caught me masturbating. He put
me over his knee and smacked my buttocks. He was in pajamas, and the slit in front of
his trousers opened widely, so that I could see his big penis and dark scrotum moving
quite near my mouth each time he raised his hand ... Ever since, I subconsciously connected the smacking of my buttocks with the view of his penis and my first sexual excitement." See Narcyz Lukianowicz, "Imaginary Sexual Partner: Visual Masturbatory
Fantasies," *Archives of General Psychiatry* 3, no. 4 (1960): 432.

flood of hormones (namely, testosterone), which quickly turns the male's reproductive system into one of those 24/7 sperm microbreweries that we encountered in chapter 3.

Since the days of Stekel, most of the causal theories about the paraphilias have been grounded in neo-Freudian psychodynamics (which emphasize the sleepless battles being waged between the conscious and the subconscious parts of our personalities).* But regardless of your take on Freud's ideas, there's no shortage of compelling case studies to support a paraphilic model of male imprinting. You'll recall, for instance, those lovers of lost limbs, the acrotomophiles. It turns out that one of the most detailed investigations of a deviant desire seeding in a fertile young mind involves a friendly amputee fetishist. It would be impossible to document every little thing that happens during a male's development to determine the event's role in crafting his lifelong attraction to, using our example here, truncated limbs, yet the study that follows comes as close as possible to doing just that.

In 1963, a team of developmental psychologists (non-Freudian ones at first) from the City University of New York began a long-term study with a group of 131 children, observing them from birth until the age of six. Funded by the National Institutes of Health, this study aimed to examine how the mother-child relationship affects the latter's emotional development and coping in a complex social world. To get at this, the investigators would unobtrusively film the pair interacting—during a feeding session when the subjects were babies, for instance, or later, while the mother and child played together—and then they'd analyze the footage to see if the child's adjustment to the outside world correlated with the quality of these caregiving exchanges. And indeed it did: children with the most secure attachments to their

---

*This psychoanalytic approach to the paraphilias traces back to Stekel's initial allegiance to Freud. Stekel's *Sexual Aberrations* served largely to pave the way for future sexologists who dissected individual cases of deviant desire using established methods.

mothers (also to their fathers, the researchers learned along the way) fared better in school and beyond.

That was the end of *that* study. But then decades later, in 1994, an entirely different team of scholars (here's where the neo-Freudians enter the picture) managed to track down most of the original child participants from the 1960s. Now in their early thirties, these people were quizzed by the psychoanalysts Henry Massie and Nathan Szajnberg on pretty much everything they'd been up to since the initial study ended. A tall order for a study, given that a quarter of a century had come and gone in the interim. Yet the results from this ambitious follow-up were fascinating in their own right. For example, the kids from those old 1960s-era home movies who'd shown the most secure attachments with their moms were now the adults in the sample that had the healthiest social relationships. For those who'd tied the knot, this included their marriages as well. But the most intriguing line of questioning for our purposes involved what was *not* recorded in the lab, namely those very personal events from these people's early childhoods that might relate to their now-adult sexuality. Most of the subjects were remarkably bland in this regard, or at least they weren't willing to divulge any particularly juicy secrets. The one exception to this was a handsome, articulate, and happily married advertising executive who reluctantly confided while sitting across from the interviewer: "The only quirk in my sexual life is that I find it very exciting to fantasize sex with a woman missing a limb." He hadn't managed to find an amputee wife, but he did find an intact one who was happy enough to simulate an amputated leg (method unknown) whenever they made love.

At first, the man had no idea how or why this peculiar desire had come to absorb his carnal consciousness. He recalled getting aroused by thinking about a woman missing a limb at the age of five or six and even innocently sharing this observation with a playmate around that time. On further reflection, though, he had an epiphany in the interview room, and Massie and Szajnberg decided that the event he was suddenly recalling was indeed the immovable "first cause" of the man's acrotomophilia.

What makes this case unique is that the careful notes collected during the attachment study all those years before allowed the man's recollections of his alleged sexual imprint to be independently verified through the details contained in his old subject files. And consistent with the time frame believed to be involved, the paraphilia took root when he was exactly five years and three months of age. It was then, in 1968, when the records show that the couple next door had moved briefly into his family's home while their own house was being renovated. As the subject was being asked about this seemingly trivial incident, it dawned on him that the attractive neighbor woman had been wearing a full leg cast and that this had transfixed his boyhood attention. One day, the husband and wife sat at the kitchen table as he'd played with his toys beneath it. He could see the woman's wounded appendage resting on the man's lap. The man sat stroking his wife's leg like a sick dog. "When's it coming off?" the boy overheard. Because the subject didn't understand that the man was referring to the *cast* rather than to the *leg*, his literal interpretation of this frightening sentiment, which happened at an age when little boys believe their fathers are going to castrate them out of jealousy (remember, these are neo-Freudians scrutinizing the incident), solidified into a paraphilia in which women amputees are eroticized heroines who've martyred one of their own limbs to save his penis. The grainy odor of the plaster cast and the image of a smooth leg stump had seared into his brain like some Proustian memory trace waiting to be enlivened by pubertal androgens.*

---

*In 1976, John Money reported on a case strikingly similar to this. Money's acrotomophilic patient recalled being around five or six years old when he'd accidentally spilled a bowl of scalding hot soup on his foot. "Get it off! Get it off!" his panicked father had shouted. Thinking that his dad meant his *foot* and not the *sock* that he was wearing at the time, this was enough, according to Money, to put the amputee fetish in motion, with the image of a removed foot allaying the boy's castration anxiety. See John Money, "Amputee Fetishism," *Maryland State Medical Journal* 25 (1976): 35–39.

Perhaps for the simple reason that it features a missing body part, acrotomophilia lends itself more readily to Freudian interpretation than other paraphilias, which explains the surplus of such stories. But I'd like to share one more, given that, in addition to being another display of male sexual imprinting, it's the most concrete manifestation of the castration complex that you'll ever hear. As a little boy, "Mr. A.," as the psychiatrists Thomas Wise and Ram Kalyanam refer to a mild-mannered forty-nine-year-old accountant, had become aroused by the medical photographs of nude, amputated women in his physician father's impeccably organized home library. After college, Mr. A. had served overseas in the military and sought out all the amputee prostitutes he could find while he was abroad, which was in fact a substantial number. He was back on American soil by the time he was twenty-eight and, a few years later, married to a woman who'd lost her leg to osteogenic carcinoma as a teenager. Yet because she was self-conscious about her own physical condition, whenever he'd fondle and kiss his wife's stump, she'd swat him away. Years into their marriage—and apparently never having had a meaningful conversation with her husband about his acrotomophilia—Mr. A.'s wife became enraged on discovering his amputee-porn collection.

With his acrotomophilic sexual outlets unplugged in this way (he wasn't going to cheat on his wife, and he could no longer pleasure himself to his preferred porn), he was presumably being deprived of the anxiety buffer that had soothed his subconscious castration fears all this time. In response to this, argued the psychoanalysts, Mr. A. regressed to the Oedipal stage by assuming the role of his jealous father, and on doing so he began fantasizing about cutting off his own penis. "This fantasy became a preferred masturbatory stimulus," they relay, "the exciting element of the idea was the actual cutting off of the penis as well as the image of the penile stump."

Now, I must say, the thought of achieving an orgasm by manipulating one's erect penis while picturing that very penis being severed is one of the most formidable cognitive hurdles I can think of. Nonetheless, Mr. A. persevered and turned this dream into a reality. What he didn't anticipate, apparently, was just how much it would hurt—and bleed. Nor does it seem to have been a particularly erotic experience for him. He ended up calling 911, and the quick-thinking EMTs were able to retrieve the source of his interminable woe and pack it on ice. A team of perplexed urologists did manage to reattach the now permanently floppy phallus, but it was thereafter reserved for delivering the kind of pleasure associated only with voiding a full bladder.

❧

The neo-Freudians have certainly dominated this psychiatric terrain, but the paraphilias have been explained by a smattering of other theories over the years as well. Unfortunately, all are just as challenging to test as the hypothesis that a middle-aged accountant with an amputee fetish just cut off his own penis in a last-ditch effort to quell his subconsciously unresolved castration anxiety from his preschool days. Throughout the 1980s, for example, the sexologist John Money laid out his "erotic love map" theory of the paraphilias, arguing that everyone is born with a default set of navigational instructions for ending up as a "normal heterosexual adult," but disruptive social events in early childhood can cause this standard route to branch out in wildly unpredictable ways. Money believed that the paraphilias emerged not from imprinting per se but from growing up in a sexually unhealthy social environment.* He cast aspersions especially on what he saw as an unforgivably puritanical United States, which

---

*Stekel might have been the first to describe them in detail, but it was Money who coined the term "acrotomophilia" after seeing several of these amputation-obsessed patients in his Johns Hopkins clinic. He "discovered" and coined many unusual paraphilias, including "symphorophilia," which is erotic arousal from staging accidents or catastrophes.

he felt was the country most likely to produce paraphilic citizens due to its great sexual restrictiveness. Money reasoned that the sexuality of the child is forced to find its way to adulthood under chokingly repressive conditions akin to a "polluted smog in biological warfare." "Like a cluster of wild mushrooms that can push through the paving of an asphalt court," he wrote (he really did like his metaphors), "[the erotic love map] cannot easily be sealed over. However, it can encounter abuses that interfere with its normal expression. The normal heterosexual play of childhood may be hampered by too much prohibition, prevention, and punishment. In that case the standard heterosexual love map does not develop properly in the brain."*

Money's ideas weren't entirely speculative. Along with his conviction that sexually oppressive societies were to blame for any such deviance, he was also among the first sexologists to observe that paraphilias are entirely unheard of in tribal societies."†
The Ila-speaking people of Africa, for example, encouraged prepubescent girls to handpick a boy "husband" at harvest time and to live as "man and wife," which included experimenting with actual intercourse. The Western world, Money argued, churns out an abundance of lascivious undesirables through its own foolish doing. I think he's right in some sense. Complaining about

---

*Money overemphasized the role of society in shaping sexuality and gender. His name will forever be ignominiously associated with the "John/Joan" case from 1965. Believing gender identity to be a product of "nurture" rather than "nature" (we now know this is a false dichotomy), Money had advised the distraught parents of an infant boy who'd lost his penis during a botched circumcision to raise their son as a girl without ever telling him the truth. This recommendation led to a major gender-identity crisis in the patient (later identified as a Canadian man from Winnipeg named David Reimer) and, ultimately, contributed to the man's suicide in 2004 at the age of thirty-eight. See Claudia Winkler, "Boy, Interrupted: A Tale of Sex, Lies, and Dr. Money," *Weekly Standard*, June 19, 2000, www.weeklystandard.com/Content/Public/Articles/000/000/011/136eioki.asp.

†It's important to note, however, that the occurrence of sexual deviance in small-scale, traditional societies has not yet been the subject of proper scientific investigation—at least of the objective, unbiased variety. So this absence of paraphilias among hunter-gatherers may in fact be more apparent than real.

our plague of sex deviants is indeed bizarre when considered in that light, given that we're only reaping what we've sown as a sexually confused, and often hostile, community. John Money pitied these erotic outliers and saw paraphiliacs as victims of their own diseased culture who are not to be blamed for their divagations from normal desire:

> [Paraphilias] are not voluntary choices. They cannot be controlled by will power. Punishment does not prevent them, and persecution does not eradicate them, but feeds them and strengthens them . . . The paraphilic person is a survivor of catastrophe . . . The tragedy that deprived the paraphiliac of heterosexual normality was the neglect and/or abuse of the rehearsal play and development of early life, and the paraphilic substitute that took its place. The triumph was that lust . . . was saved from total wreckage by being transferred to some other less prevented and at the time less censored, but paraphilic rehearsal. It was a hollow triumph, alas, for in later years, it would bring more tragedy, as a paraphilia so often does.

Money's theory of the diverted "erotic love map" relates to more recent efforts to explain paraphilic origins through the so-called Zeigarnik effect. Whether you're a paraphiliac or not, you've had your fair share of encounters with this irritating psychological experience. The Zeigarnik effect refers to our brain's relentless search efforts in the wake of some interrupted goal or activity. As a humdrum example of this feeling of getting psychologically stuck on something, imagine going out to the movies and seeing that new flick that everyone's been raving about; you're totally absorbed in the film, but just as it gets to the part where everything is about to come together in the plotline, the theater's fire alarm goes off. You'd be experiencing the Zeigarnik effect out in the parking lot while waiting for the manager to confirm it's just a false alarm, with all your frustrating thoughts about how the story is going to end chafing away at your brain. It's a bothersome feeling.

When the idea of the Zeigarnik effect is extended to the paraphilias, the theory is that if a child's inchoate sexual feelings somehow get stimulated (which is natural), his brain will launch into a problem-solving mode if his curiosity fails to be satisfied. It's easy to see how this could happen. As a consequence of our enshrouding sex in mystery and the forbidden, children sense a conspiracy of silence regarding unspeakable acts and untouchable parts, and sex, ironically, is therefore made all the more salient and attention-getting to them. With adults seemingly hiding something so gravely serious and ever so important, the enigma only widens. Wherever they look, they're reminded of this shadowy unsolved "problem" that nobody talks about and which, therefore, keeps nipping away at them. (Interestingly enough, many of the original sexologists, including Havelock Ellis, believed that those children who are the most naturally inquisitive and bright—the intuitive problem solvers—are also the ones most susceptible to developing paraphilias.)

What to make of this, though? Most of us see four- to nine-year-olds as asexual, but if this is indeed when irreversible male sexual imprinting occurs, our denial that children have any capacity for sexual feelings may actually increase the odds of us brewing deviant little darlings. After all, an eroticized Zeigarnik effect is more likely to occur given that the child's ostensibly "sexless" world is, in reality, to *them* over-sexualized in cryptic ways. Perhaps when paired with a genetic predisposition for sexual imprinting, some little boys' nagging cognitive efforts to solve the grand riddle therein will seep into parts of their environment that wouldn't otherwise be sexual. These misplaced sex cues can inveigle deep into brain networks linked to their arousal, and voilà, a paraphilia is born. In other words, we might want to reconsider the "appropriate" age at which we, as a society, provide clear sex education to children.

More than half a century ago, Alfred Kinsey lamented the state of the paraphiliac and such other "perverts" and deviants, reflecting specifically on his society's delusion of having made much moral progress when it comes to thinking critically about human sexuality. "We no longer burn those who were possessed of devils at birth," he wrote, "but we still hold the individual responsible for the spirits that take possession of him after birth." Unfortunately, that sad social commentary is still true for us today. But I'm convinced it really doesn't have to be so. The more we understand about each other, the less we have to fear.

FIVE

IT'S SUBJECTIVE, MY DEAR

> While we are dealing here with pain, it is a pain the masochist is capable of transforming into pleasure; a suffering which he, by some secret alchemy he alone possesses, can turn into pure joy.
>
> —Jean Paulhan, Preface to Pauline Réage's
> *The Story of O* (1954)

If you were to tell me the first thing that comes to your mind when I say "S&M," there's a decent chance that it would be the *Fifty Shades* trilogy. Or perhaps you'd describe some sensual scene lit by flickering candles: an ice cube to a perky nipple, wax dripping onto a shivering naked abdomen, a guillotine mask, a pony harness, a strangled scrotum, and a spring-loaded mouth gag . . . well, now you're just getting carried away, you perv. In any event, this dominant-submissive (or "master-slave") dynamic is more common than you may be aware. Around 11 percent of us, irrespective of our gender, have had an experience with sadomasochism. Five percent of men and 7 percent of women also report regularly including "verbal humiliation" in their erotic repertoire. And that's just those who'd ever admit to such things on a survey.

Most so-called sadists in the S&M community aren't really

that sadistic; nor are most "masochists" willing to commit *too* much to their masochistic roles. Emphasizing the relative harmlessness on both sides, John Money referred to the average bondage-and-discipline fan as a "velvet dragon." Although sadism is included in the *DSM-5* as one of the harmful paraphilias, clinicians aren't going to favor a mental illness diagnosis for some theatrically inspired woman who likes to occasionally tie up her "prissy little bitch" of a consenting husband and sink her teeth into his buttocks. These things might smart and induce some wincing, perhaps a few tears, but they're not exactly going to lead to reconstructive surgery or to a major medical claim. Rather, to determine if the patient's kinkiness qualifies her as a genuine sexual sadist, psychiatrists look for two things: the pain she inflicts must be real rather than playful, and she derives her most intense pleasure from the suffering of a nonconsenting other. Usually, such a person—that's to say, a paraphilic sadist—will only come to our attention after she's committed a serious crime, with the courts ordering her to undergo a psychiatric evaluation.*

That's sensible enough, since it hurts so good except when it *doesn't*. For horrendous incidents in which deranged individuals attack innocent victims, pain and nonconsensual sex almost always go together. When these two criteria are applied to S&M outliers far off the beaten path from the rest of that community, however, there's often plenty of real pain and suffering, heaps of it even. Yet the receiving party has also consented to the sex act. Don't forget that all-important fact about the subjectivity of harm and how it varies from person to person. Cultural differences come into play here as well.† And with an extreme mas-

*The example used here is of a female paraphilic sadist. But as we learned in the previous chapter, males are more likely to be "true" sexual sadists, especially of the criminal variety.
†For example, in the nineteenth century, Apinajé women in Brazil were known to bite off their male lovers' eyebrows and spit them out while having sex. The husbands of

ochist the issue of harm can get especially tricky. Screw on your forensic cap, for instance, and consider how you'd analyze the particularly gruesome case of Der Metzgermeister (the Master Butcher) of Rotenburg, Germany.

Before the more recent case of the Cannibal Cop (in which the New York City police officer Gilberto Valle was found guilty of planning to kidnap, murder, and eat a woman—or eat and then murder, the correct order of those last two verbs isn't clear), there was the tale of Armin Meiwes. That name might dimly light up in your frontal lobes as a distant headliner best forgotten, but to remind you—and apologies for doing so— Meiwes, a computer repairman by trade, was the German cannibalistic sexual sadist who, in the spring of 2001, found a willingly edible sexual masochist for himself named Bernd Jürgen Brandes. "Looking for a well-built 18- to 30-year-old to be slaughtered and then consumed," read Meiwes's personal ad on the Cannibal Café website (which, you'll probably be glad to know, is a website no more). And Brandes, it seems, was looking to be thoroughly digested. That old psychopath Cupid couldn't have arranged for a more tragic crossing of these lovers' paths. (Oh right, where's my head? After our terminology lesson in chapter 4, you're probably starving to be told the technical name for this cannibalism paraphilia. It's "vorarephilia," from the Latin *vorare*, "to swallow or devour.")

From the video footage—the whole awful thing was taped from start to finish—it's apparent that Meiwes (the eat*er*) didn't

---

Trukese women in the Caroline Islands could anticipate a finger being poked sharply into their ear canals by their highly aroused wives. And prior to the invasion of the more conservative Muslim rulers, sex among the early Hindus of India was a refined form of combat. As the psychiatrist Dinesh Bhugra describes it: "The male attacks, the woman resists and, amid the subtle interplay of advance, retreat, assault and defence, the desires are built up." Bhugra explains that Hindu consciousness was enhanced during sex by first dulling the physical sensations through sadistic acts. See Dinesh Bhugra, "Disturbances in Objects of Desire: Cross-Cultural Issues," *Sexual and Relationship Therapy* 15, no. 1 (2000): 69.

coerce Brandes (the eat*ee*) at all. If anything, the coercion was done by the masochist, not the sadist. Brandes even begged a hesitating Meiwes to bite off his penis. There's no need to hover over all the nasty details, so let's fast-forward to the kitchen scene, where Meiwes has just sautéed the chomped-off organ in a pan with wine and garlic and the pair is now amicably dining on it *together*. (Meiwes would later describe it as "chewy.") Forward still, past that unmentionable bit and so on . . . and then finally to the part when Brandes's salted carcass is hanging on a butcher's rack in the purposefully built slaughter room of this house of horrors, where forty-four choice pounds of him would slowly disappear down Meiwes's esophagus. I'm by no means suggesting that both the *S* and the *M* parts of this equation weren't disturbed men. You don't need to be a mental health expert to see that. Yet when we try to apply the *DSM-5*'s criteria of pain and non-consent, there's some tension here, since Meiwes had explicitly sought out a consenting adult partner. And when you combine consent with a masochist's apparent death wish, well, you can see how a forensic psychiatrist's job can be daunting. Diagnosing the only remaining member of this carnal coupling becomes, shall we say, "philosophically problematic," given the victim's extraordinarily unique subjectivity.*

Attempts to objectify pain and consent for a diagnosis of sadism also run into trouble at the other end of the ouch scale. In this case, it's not consent that's the issue (the other person clearly doesn't want to be doing what the sadist is enjoying); rather, there are some sex acts that don't appear to our own eyes to be "painful," at least as we usually understand that term. In fact, they can very well escape our notice altogether as being sexually motivated, but from the actor's point of view, they're maliciously

---

*In the end, Meiwes was charged with murder and sentenced to life in prison. The judgment was rendered with considerable controversy since both parties were consenting adults. The case is liable to inspire drunken debates for at least another century.

so. Take tickling, for instance. Despite peals of laughter and the ear-to-ear smiles associated with tickling, it can be a very unpleasant experience for those on the receiving end of a well-placed feather or incessant forefinger. Consensual tickling isn't unusual for those into S&M, and usually it's quite harmless. But when it's done mercilessly against someone's will, tickling is no less than torture.

When you combine a sadistic personality with the paraphilia of "titillagnia" (erotic arousal from tickling), the result isn't anything to laugh about at all. I submit for your consideration a patient described in 1947 by the psychiatrist Emil Gutheil. The man was a married, prominent thirty-nine-year-old lawyer living in New York City (funny how so many of these deviant sex cases are based in the Big Apple) who just happened to have an ineradicable sexual urge to tickle people. This wasn't some innocuous velvet dragon, however. Instead, his most frequent masturbation fantasies involved tickling a person to *death*, or at least to the point that the person would foam at the mouth and lose consciousness. (This man's sexual sadism may have given the case a unique spin, but he certainly wasn't the first to tickle for malevolent reasons. In ancient China, the courts of the Han dynasty brutalized the nobility by tickling them, since it left no marks on these high-profile figures when they'd be seen in public. Centuries later, a woman in Vienna filed for divorce against her husband, who for years had been tickling her as a form of spousal abuse for precisely the same reason.*)

Consistent with the theory of male sexual imprinting, Gutheil believed that the patient's desires stemmed from him growing incidentally aroused as a seven-year-old boy while being overpowered and tickled by his older brother. Now an adult, the man found intercourse repulsive, and he'd been forcing his poor wife

*There's also the old story warped by time and of questionable veracity about a tailor from Salzburg who'd allegedly murdered seven of his wives by tickling them to death.

to indulge his vicious tickles. The psychiatrist notes how the sadistic patient was sometimes able to find prostitutes who'd allow him to tie them up and tickle them. "But this type of prostitute was so expensive that he could not afford them for any length of time," writes the psychiatrist. The man also had a history of paying young boys to tickle each other unremittingly, masturbating on the sidelines while watching them doing so and thereby re-creating the childhood scene with his brother but this time in the role of voyeur.

It's kind of a pity that this man lived when he did. Today he'd surely have found a more suitable "knismolagniac" (the tickling masochist to the titillagniac's sadist) or perhaps even the more rarefied "pteronophiliac" (those who obtain their most intense gratification from being tickled with feathers) somewhere online. (If a computer repairman can find a person happy to be swallowed alive these days, an attorney can certainly find someone willing to be tickle-tortured.) Had the patient enjoyed his sexuality with one of those complementary paraphiliacs, the harmfulness of his actions would have been less than they probably were for his wife, prostitutes, and children.

Sadism isn't the only paraphilic category for which the question of harm can get murky for psychiatrists as well as for anyone who has ever contemplated another person's unusual sex life. As the lovely Kate Upton reminded us earlier, a universally objective reality simply doesn't exist in the present domain; what's harmful to me isn't necessarily harmful to you, and vice versa. It will change as soon as I put this comma right *here*, but as of this very moment there are exactly 7,088,343,858 people on the planet. If all but one of these individuals were to experience harm in exactly the same way from a certain sex act, that solitary person is nevertheless just as right (or just as wrong) as all the others combined. This is because there's no "correct" way to experience a sex act, only individual differences in subjective realities. It may be a moot point, since it's not logic that guides culture but instead sheer social mass shouldering into it with brute force, but nonetheless 7,088,343,857 shared subjective realities do not add up to a single objective fact. What was harmful to *them* was not harmful to *him*, and that, as they say, is that. Or to rephrase: one person's horror story is another's erotica. And I'm quite sure our vorarephile Bernd Jürgen Brandes would tell you so too, if only he were still around.

That's all well and good, you might point out, but there are issues here far more serious than two people disagreeing over whether, say, getting tickled by feathers is sexy or cruel. Take pedophilia, for instance. Is harmfulness "subjective" with that, too? Well, *yes* and *no*. This is a dark and treacherous path that we're about to embark on, so it's all the more important for us to be guided by the light of clear thinking. There's no doubt whatsoever that children who are sexually abused are often irreparably damaged— and not only psychologically but, sadly, physically as well. Denying that children can face grievous, debilitating, and permanent harm from being violated by adults in this way is just asinine. It flies in the face of scientific data showing such trauma is all too real.

Yet research has also revealed that not every child who has a sexual encounter with an adult is traumatized. In the next chapter, we'll get all of our terms straight regarding the "erotic age orientations"—there are at least five of them—but one important thing to highlight here is that when it comes to sex, there's a world of difference between a six-year-old and a sixteen-year-old. (The term "pedophilia" has become so misused that it's now difficult to reclaim its proper scientific meaning from the vernacular, but this area of what John Money called the "chronophilias" gets fairly complex.) A six-year-old child may subjectively experience little, if any, harm at the time of being molested by an adult, but that doesn't mean significant damage hasn't been done. As the child grows older, his or her interpretation of the experience on coming to understand what really happened may become increasingly traumatizing. One way to think about it is that the sexually abused child has essentially been implanted with a psychological time bomb that may or may not go off down the road. There's a marginal chance that it won't detonate at all, but if it does, it's often catastrophic. This can also be the case with teens (many of us have regrets about being taken advantage of when we were young and naive, sexually or otherwise), but for a sixteen-year-old who has hormonally fueled desires that are just as intense as those of the adult he's having an encounter with, it's a very different can of emotional worms.

The more controversial "yes" part of my response to the question of harm in such sexual encounters with adults being subjective, therefore, refers to the fact that for some individuals looking back on their childhood or adolescent experiences, the event was not harmful *to them*. For whatever reason, their bombs never went off. Whether or not we have a hard time understanding their perspective, there are indeed people out there who feel this way. (And some who, believe it or not, even view their sexual experience with an adult as a positive moment in their lives.) The "no" part of my response, by contrast, relates to the fact that the child's sexual subjectivity inevitably changes as he or she grows older. To say that a six-year-old who doesn't understand what's happening to her isn't being harmed because she appears to be just fine otherwise, or even that she "likes it" (as child molesters sometimes claim), is to miss the point entirely of these delayed, and potentially devastating, emotional injuries that may affect her later. If we were to follow up with this little girl in twenty years, we might find a woman damaged beyond repair by the memories of the very events that were inconsequential to her at the age of six. In other words, the bomb has gone off.

So although harm may not be inevitable in every case, what *is* inevitable is the very high risk of shattering a child's life for the sake of one's own immediate sexual gratification. (And note that this is true regardless of cultural differences in symbolic disgust: if the pain is real to the individual, then it's real in terms of its harmful effects on them.) When it comes to adolescents' sexual experiences with adults, where the distance between the subjective present of the child and his or her subjective future has grown narrower, some researchers believe that the damage isn't as significant as it's often assumed to be. Bruce Rind, for example, an expert on the study of "intergenerational sexuality," set off some explosions of his own in the late 1990s when he published a set of highly contentious (to put it mildly) findings to this effect. The best predictor of subjective harm—past, present, and future—he found, is the minor's lack of *consent*. Obviously, there's consent in the legal, underage sense of the term, but there's also consent as a mental state (basically, the feeling of wanting to do something) that occurs regardless of age. Rind was more interested in this latter, psychological meaning.

In 1998, he and his coauthors, Philip Tromovitch and Robert Bauserman, managed to put the American Psychological Association (or "the other APA," which comprises not psychiatrists but academic, research-oriented clinical psychologists) in the rather awkward position of having to publicly acknowledge that not every incident of an adult having sex with a juvenile is harmful. Actually, "awkward" may not be entirely the right way to describe it: it was more that the APA locked horns with every wrathful, powerful politician in the country for defending this very politically incorrect view. It all started when Rind and his colleagues published a study in the association's flagship journal, *Psychological Bulletin*. The authors argued that it makes little sense to refer to something as "child sex abuse" if, as an adult, the individual doesn't personally feel harmed and if his or her harm can't be detected by any known empirical measures.

The most delicate issue for the APA was that this wasn't just the authors' personal and controversial opinion, but a statement based on scientific findings. Rind's study was a meta-analysis of previously published data on the sexual histories of a whopping 35,303 college students from around the world. One thing that set his project apart from most others in the field of child sex abuse was that the data being evaluated weren't from clinical samples of adults who'd sought help for ongoing problems stemming from their being raped, molested, or otherwise sexually exploited as children or teenagers, but from random samples of college students. And what Rind and his coauthors found by using a large *nonclinical* sample was that the majority of those people who reported having had consensual encounters with adults as minors were, at the time of testing, no more likely to have pervasive psychological problems than those who hadn't.

Now, it's extremely important to bear in mind that for the most part, these mentally healthy individuals weren't those who'd been subjected to terrible abuses as children. More often they were those who, as adolescents, had consensually (again, in the psychological sense of that term) "fooled around" in various ways with someone on the other side of the legal line (wherever that line had been drawn at the time, since, remember, it was an international sample and, as we'll be examining in detail shortly, the legal age of consent varies by country). Beyond that weird and unwanted guinea-pig incident with the hebephilic pygophile (which wasn't consensual even after the fourth groping of my skeletal backside), I never had any such experience with an adult while growing up, so it's hard for me to know for certain. But speaking only for myself, I suspect that had a very adult Mr. April 1991 stepped out of the calendar that hung in my sister's dorm room that year, ripped, hairless legs beaded with river water, his unkempt hair plastered against his forehead and glistening clavicles, wearing nothing but wild green eyes and a tapioca-colored fishnet loincloth (my *vague* recollection), and proceeded to gently guide me on the ways of man-on-man love, the last thing that my fifteen-year-old self would have felt was harmed. I only wish I had such a memory to look back on now.*

You might think that Rind and his fellow investigators had a dark agenda of some sort (and certainly many critics have accused them of such), but in fact they were exceedingly careful to note that their findings shouldn't be used to implement any policy changes to existing laws concerning underage sex. Rather, their intention was only to call into question some widely held assumptions about the universal harmfulness of such developmental experiences. But then the conservative radio talk-show host Dr. Laura Schlessinger (whose Ph.D., incidentally, is in physiology, not psychology, not that she's ever let that get in the way of giving mental health advice with a side of Old Testament gloom) somehow got wind of it and proceeded to stir up a bit of a hornet's nest by complaining about Rind's "junk science" and pedophilia apologia to listeners of her nationally syndicated program. That incendiary development, combined with NAMBLA's (the notorious North American Man/Boy Love Association) gushing over Rind's findings on its website, quickly escalated into outright political chaos.

Before long, some members of Congress learned of the study (whether that was from being regular listeners of Schlessinger's show or from being loyal followers of NAMBLA is hard to say). By the spring of 1999, Alaska, California, Illinois, Oklahoma, Louisiana, and Pennsylvania had passed official resolutions executively condemning Rind's scientific findings. On July 12 of that same year, in an incident without precedent in the history of psychology, the U.S. House of Representatives convened to establish Resolution 107, in which 355 members of the House (all of whom were even less qualified than Dr. Laura Schlessinger to

---

*On the other hand, had he forced himself on me, I could be a quivering pile of jelly right now. But the point is that even when it comes to illegal sex with a minor, the likelihood of psychological harm is reduced by the minor's consenting mental state.

evaluate a study in psychological science) legislatively lampooned Rind's empirical work by declaring—data be damned—that any and all sexual relations between minors and adults were categorically abusive and harmful. Just a few weeks later, the Senate passed this anti-Rind resolution by a unanimous "Hear! Hear!" voice-vote margin of 100 to 0. (No senator would dare to go against the grain and kiss any future political aspirations goodbye by being known as the one pervert voting in favor of Rind's findings.)*

The advisory board members of the APA, meanwhile, butted heads with these suits in Washington, D.C., over the fallout. After all, the former were in the business of science, not moralizing. They staunchly defended the *Psychological Bulletin* editor's decision to publish the study in their prestigious outlet. Moreover, the study had been carefully vetted—and ultimately recommended for acceptance—by other experts in the field. The Rind data remain polarizing, with some researchers still questioning his methods and motives, and others, in turn, questioning *their* motives for questioning his. Given the brouhaha, scholars on both sides of the debate left things for a while on a "Let's just agree to disagree" note (which essentially meant cursing each other under their breath and taking a break from duking it out in public). But in 2006, the psychologist Heather Ulrich replicated the 1998 findings, concluding cautiously that the presumption of universal harm from juveniles having a sexual encounter with an adult is too simplistic to account for the variance in people's subjective interpretations of their own life experiences.

*One of the few politicians to abstain from voting on the study was Representative Brian Baird, a Democrat from the state of Washington who boasted a Ph.D. in clinical psychology. Baird wanted to go on record as saying that out of the 535 members of the House and Senate, only 10 had actually bothered to read Rind's article. See Brian N. Baird, "Politics, Operant Conditioning, Galileo, and the American Psychological Association's Response to Rind et al.," *American Psychologist* 57, no. 3 (2002): 189–92.

•

Now, ostensibly, there's no good way for one to lean on this hot-button issue. If you conclude from Rind's findings that having sex with minors is "sometimes okay," then to many people you'll sound like an advocate of child molestation. Yet if you still get red in the face and believe that anyone having sex with a minor is manipulating an innocent child, then you're glossing over all the gray areas. Much of the disagreement, I think, stems from our failure to define our terms. When, exactly, does "childhood innocence" end? Most of us have some vague sense of once having had it—or at least something like it—but how does one quantify or even standardize such an abstract construct? At what precise moment in time, for instance, did you lose yours? Perhaps you never had it, or perhaps you never lost it.

Few of us are so naive as to believe that it happens at the stroke of midnight dividing childhood from legal adulthood, any-way, especially given that such a line is culturally arbitrary. There are a lot of uncomfortable philosophical problems with age-of-consent laws that continue to lead to people, including teenagers themselves who have sex with someone just three or four years younger, being treated unfairly in our society. It's only when it comes to sex that on some mandated calendar day the legal con-cept of "consent" changes so abruptly from being a chronological state to a mental state. Having sex with a person before the bell tolls on that day, even if it's the minor who makes the first move and you're but the target of his or her passionate underage desires, will change you abruptly into a criminal sex offender. Whatever was going on inside the minor's head is usually seen as inconse-quential. In other legal contexts, however, minors are tried as adults precisely *because* of what was going on inside their heads at the time. A sixteen-year-old boy who rapes a woman after she rejects his solicitous advances would typically be punished as an adult. But if a woman encourages a solicitous sixteen-year-old boy after he tries to kiss her, he'd be regarded by criminal prosecutors as a child victim. In other words, legally, the minds of minors matter only when they've caused adults sexual harm; their mental states are inadmissible in court when they cause adults sexual pleasure. I don't know about you, but to me there's an unsettling tension in the logic between these contrasting scenarios.

Age-of-consent laws are quite admirably meant to protect adolescents from being sexually exploited by adults (and there are, sad to say, plenty of the latter out there for parents to be concerned about). But there are problems with a hard-line approach to this emotional immaturity argument as well. One might be stigmatized for doing so, but it's perfectly lawful to have sex with consenting adults who have the intellectual and emotional capacity of an underage child. To take a rather extreme example, the average mental age of an adult with Down syndrome is eight, yet unlike having sex with a seventeen-year-old equipped with a three-digit IQ, being with someone with this or any other developmental delay isn't a crime, so long as the person is eighteen years of age or older and "consents." So if we're really trying to protect the vulnerable from sexual harm due to their mental immaturity, then using *chronological* age, rather than *mental* age, to draw the legal line seems a somewhat odd way to go about it. I can assure you after an early dating mishap with one particular—to put it both kindly and mildly—intellectually blunt grown man, these are often orthogonal measures.

Recent work by the cognitive neuroscientist Sarah-Jayne Blakemore shows in fact that there's no hard-and-fast threshold at which a person crosses over into a clear brain-based psychological adulthood. The prefrontal cortex, arguably the neuroanatomical region most relevant to sexual decision making due to its executive role in long-term planning abilities, empathy, and social awareness, doesn't stop growing until we're in our mid-thirties. For some, it's still developing well into the fifth decade of life. Those who want to have sex with minors often cite in their defense the cultural arbitrariness of ages of consent. But they're at least right about that, and these numbers are in perpetual flux even *within* cultures.

The first such age-restricting statute, Westminster I, appeared about seven centuries before the popular TV series *To Catch a Predator* first aired, in the year of our Lord 1275, under the heading of a broader rape law in England. According to this new legislation, any man who dared to "ravish" a "maiden within age," with or without her consent, was guilty of a misdemeanor. English legal scholars interpret the phrase "within age" to mean the age of marriage, which at the time was twelve. Had the man been married to this same twelve-year-old girl, in other words, this age-based rape statute wouldn't have applied. In the centuries that followed, similar edicts meant to protect children (namely girls) from sexual abuse or exploitation by adults (namely men) started to dust the globe. And wide variations in the age of consent are written across this historical landscape.

In the sixteenth century, for example, the North American colonies adopted from Britain the age of ten as the appropriate cutoff, and this remained in the formal legislatures of thirty-seven U.S. states until long after the Civil War. Of the other existing states in the 1880s, only nine had by then decided that the "advanced age" of twelve was probably a more reasonable number. (One state, Delaware, had even lowered its cutoff to a mind-boggling *seven*.) Only in the late nineteenth century was the age of consent raised to sixteen throughout most of America, a concession by the social reformers who had spearheaded the campaign and had initially sought to have it changed nationwide to eighteen, which they'd largely accomplish by 1920. Some in the growing feminist movement even hoped to raise it to twenty-one, the age at which women could legally inherit property.*

---

*Today, each state has its own numerous and complex series of clauses dealing with factors such as age differences between parties and the nature of specific sexual acts, but general ages of consent presently range from fifteen (in Colorado only) to eighteen.

Tidal changes were happening with Europe's age-of-consent laws over this time span as well. With the Enlightenment really coming into bloom in the eighteenth century, the philosopher Jean-Jacques Rousseau's influential ideas on childhood spread throughout France and beyond. His portrayal of children as social tabulae rasae, or "blank slates," lasted well into the Napoleonic era and replaced the archaic notion of kids as adults in miniature. Inspired by Rousseau, public sentiment came to hold that children were intrinsically pure and became tainted only by the corrupting influence of society. Rousseau also marks the dawn of developmental psychology and the implementation of age-segregated education. Children and teenagers were now seen as having qualitatively different kinds of minds from grown-ups, marching through what Rousseau believed was a universal pattern of cognitive and emotional stages. (Development wasn't just a matter of acquiring more and more facts, in other words, but the very way in which one processes information and perceives the world changes over time as well.) Yet even at the peak of this radical new moral Enlightenment, the sexual readiness of children was, strangely enough, apparently seen as a separate issue. The age of consent in France during this whole time was a mere eleven, getting bumped up to thirteen only in the year 1863.

By the late nineteenth century, thirteen had also been chosen as the carnal threshold in other European nations, including Portugal, Switzerland, Spain, and Denmark. Today, Spain is the only country in the region to keep thirteen as its age of consent, with other nations variously lifting theirs to fourteen, fifteen, or sixteen, at most. Deplorable tales of child prostitution during the Industrial Revolution spurred moral reformers in England and Wales, meanwhile, to raise the age of consent across the British Isles from thirteen to sixteen, a social cause to combat child exploitation that had also reverberated in the American age-of-consent campaign mentioned earlier.

Similarly wobbly views on sex and adolescents—or rather sex

with adolescents—are on profligate historical display elsewhere. It goes in the opposite direction, too. The age of consent in 1920s Chile was twenty, but now it's sixteen. A century ago in Italy, it was sixteen, too. But today it's fourteen there. Overall, studying the numbers contained in even the most contemporary international age-of-consent table will give you the impression that you're looking at a flurry of seemingly random digits between twelve and twenty-one (a sizable range): it's thirteen in Argentina, eighteen in Turkey, sixteen in Canada, twelve in Mexico, twenty in Tunisia, sixteen in Western Australia, fifteen in Sweden, and so on. "More than 800 years after the first recorded age of consent laws," writes the historian Stephen Robertson, "the one constant is the lack of consistency." Just as when we're assessing religions with conflicting theologies, we can draw only two possible conclusions from Robertson's observation: either some societies have the one true age of consent and every other has therefore got it wrong, or any given society's age of consent is based on what its citizens have simply chosen to believe about human sexuality and psychological development. And similar to what any objective analysis of competing religious beliefs would force us to conclude, there's no evidence that the former is the case for cultural variations in age of consent laws (that there is "one true age") and every reason for us to conclude the latter is in fact what we're dealing with.

～

In the context of this somewhat scandalous discussion, it's important to remember that earlier distinction in moral logic, the one that assigns a different harm value to thought and action. Even when it comes to "true pedophiles" (those who are actually attracted to prepubescent children, not adolescents), thoughts alone are abstractions that—as unpalatable as they may be—cannot rape, molest, touch, batter, or bruise. If contained in the mind, desires are intrinsically harmless to children; only when they're driven into action can harm occur. But there's also that rather salient personal distress qualification for the paraphilias in the *DSM-5* that we went over in the last chapter, and so regardless of whether one's sexual deviancy is harmful to anyone else, it could still be causing subjective harm to the unfortunate paraphiliac in whose brain it burns. In the world of nonsexual mental illness, for instance, the voice in a schizophrenic's head may be distressing to that person whether or not the auditory hallucinations are responded to out loud for anyone else to hear. Having to listen to obnoxious personalities belittling you or egging you on all day long isn't the most pleasant way to live one's life, after all. Similarly, having a paraphilia can be a living hell even if it's kept under cranial lock and key. A paraphilia is a way of seeing the world through a singular sexual lens, and this lens can't be easily adjusted, repaired, or even, in the absence of a lobotomy anyway, broken. The paraphiliac's deviant desires are far less treatable than the voices jeering a schizophrenic. (If you're a schizophrenic paraphiliac, I only wish there were a God around to have mercy on your tortured soul.)

As we saw in the previous chapter concerning gender differences in sexual imprinting and erotic plasticity, once a male's desires calcify into such a discrete pattern of arousal, it's a permanent affair. Somehow or another, the paraphiliac must come to accept that, to live with the reality of his "socially inappropriate" psychological existence. A good therapist might be able to correct any obvious bad habits and decision making, or the patient's overall sex drive can be watered down with libido-crushing medication (such as Depo-Provera), but the unique design of his erotic taste buds remains as deeply etched in his neurons as are the fingerprints on his hands. Quibbling over whether paraphilias should be seen as "sexual orientations" and be recognized as such for political or social reasons is entirely irrelevant in this sense. Of *course* they're sexual orientations; a paraphiliac's brain orients him to an atypical erotic target (or activity), just as other people's brains orient them to the normal suspects.

And pedophiles aren't the only ones likely to lead morbidly troubled inner lives. When you're oriented to erotic targets that can fit in the palm of your hand, for example, and that most people would sooner step on than screw, that's also a pretty tough row to hoe. These are the so-called formicophiles (from *formīca*, Greek for "ant"), and the difficulty they face with their unusual paraphilia is exemplified in a 1987 case study by the Sri Lankan psychiatrist Ratnin Dewaraja. The psychiatrist explains how an introverted young man entered his clinic seeking treatment for what the patient had called "my disgusting habit." I think we crossed that line long ago and you aren't likely to be shocked by anything at this stage, but a formicophile's most intense sexual urges involve placing small creatures like snails, frogs, ants, or roaches around his erotic zones (genitals, anus, and nipples, usually), then pleasuring himself to the tiny nibbling mandibles, or perhaps the cold slime trail forming behind a slug as it makes its arduous hike across the twin peaks of his testes. "He was depressed, unemployed, had no friends, and most of the time," Dewaraja tells us of this bleary-eyed man torn straight from the pages of a Tim Burton film script (if Tim Burton produced niche porn, that is), was "preoccupied with collecting [such specimens]." After a year of counseling, the patient had managed to reduce his formicophilic masturbation sessions to just once a week, down noticeably from the three to four weekly bouts at the start of his visits. "He was [now] engaging in social interactions with women on a regular basis," Dewaraja writes optimistically, but then adds tellingly "[he] had not yet experienced coitus."

It's clear enough that the formicophile in this story had a paraphilia that he didn't want, given that it was interfering with his life in negative ways. It was subjectively harmful to him because it was associated with his own personal distress. But note that his formicophilia isn't the sole cause of this distress. If he'd grown up in a culture that revered formicophiles as reincarnated deities, for instance, his experience would likely be entirely different.* In other words, a paraphiliac's level of distress is usually correlated with the extent to which his society demonizes, ridicules, or shames his form of deviance. Given the objects of their affection, necrophiles are more likely to be demonized than are transvestic fetishists, who in turn are more likely to be ridiculed (mostly in bad British comedy skits). The most shamed (and feared) paraphiliacs today are the pedophiles. A lifetime of having to continually defend, rationalize, or hide for dear life any such unwelcome paraphilia in a society that not only doesn't understand it but doesn't *want* to understand it is obviously going to serve as a petri dish for sprouting personal distress symptoms. These people aren't living their lives in the closet; they're eternally hunkered down in a panic room and chewing away nervously at their nails.

Homosexuality is no longer regarded as a paraphilia, but as a gay man who tried to pass as straight for the first twenty or so years of his life, I can assure you that hiding one's "true nature" from the world is absolutely exhausting. Here's an exercise in the hypothetical that may be helpful for those of you who fall more along the far vanilla side of the Neapolitan ice cream erotic equation. Let's say you've been placed in a witness protection program and you suddenly have to create a new identity of being *gay*, which is the most vital part of your cover. You must move all

---

*This is the same chicken-and-egg problem that we saw in chapter 3 involving the notion of personal distress for ego-dystonic homosexuality and hypersexual disorder. Is it the person's sexuality that's responsible for his or her personal distress in these cases, or is his or her personal distress the result of living in a society that refuses to accept even harmless expressions of sexual diversity? With the exception perhaps of the creepy crawlies getting accidentally smashed by a vigorous hand, formicophilia certainly isn't harmful to others. When society rejects the self, the self rejects the self.

alone to a place where nobody knows you, and you must convince everyone you meet, for your own safety and for the safety of those you care most about, that you're 100 percent homosexual. Now, don't try *too* hard to appear gay, because you'll give yourself away, so be stereotypical but not *too* stereotypical, yet don't ever let your guard down either, since some people will try to trick you into revealing the truth by being "understanding," and it's hard to know if they actually do know, too, so err on the side of caution and assume they don't. Watch what you say, where your eyes go, what you do in your spare time, whom you're seen with, and careful, now, no matter how close you get to someone in this new life of yours, no one must ever discover that you're really a heterosexual. All that you know and hold dear—and I can't emphasize this part enough—hangs in the balance. Whatever you do, and in fact you better make this your mantra, *don't be yourself.*

If living under such intense social conditions for the next twenty, forty, sixty, or even eighty years wouldn't do a number on your nerves (and by that I mean cause "personal distress"), then you're simply not human. Yet this is exactly how many people today live their entire lives.* It's also true, however, that many deviants aren't bothered at all over their minority sexual orientation. This is because the vast majority of the paraphilias are either so rare (such as "lithophilia," an attraction to stones and gravel), so common (such as foot fetishism), or so trivial (such as "katoptronophilia," a need for sex in front of mirrors) that passing as normal isn't so much a hide-for-dear-life sort of problem

---

*A pair of sociologists once examined the discussion threads of pedophile chat rooms and noted how many topics involved tactics of "passing" in society. "Look at the boys with their mothers next to them," advised one when asked how not to draw suspicion. "If a friend notices that your attention is elsewhere, just comment on the mother." See Thomas J. Holt, Kristie R. Blevins, and Natasha Burkert, "Considering the Pedophile Subculture Online," *Sexual Abuse: A Journal of Research and Treatment* 22, no. 1 (2010): 18.

for these people as it is a "nobody-would-believe-me-anyway"
problem, a "you-and-everyone-else-I-know" problem, or a "so-
that's-your-big-secret-after-all-this?" problem.

Still other sexual orientations are altogether impossible to hide
from society's prying eyes, and this constant scrutiny can be incredibly painful for an individual who doesn't particularly want
any attention. Some people, that is, have no choice but to wear
their deviancy on their sleeves, quite literally when it comes to
"cross-dressers" in all their varied forms. There are many subcat-
egories of people who wear the clothes of the opposite sex, with
each group having a different motivation for doing so. Some of
these individuals clearly have sexual motivations, whereas others
have motivations that are anything but erotic. And some, well,
it's not entirely clear to scientists *what* their motivations are, and
for reasons we'll see shortly, that uncertainty continues to be the
source of considerable conflict. But whatever reasons one has for
needing to transform into the opposite sex, hiding for dear life to
minimize feelings of personal distress clearly isn't an option. Instead, without the right bone structure and a good surgeon, his
or her difference is exposed for all the world to see.

That's not necessarily always such a bad thing, mind you, especially for those who want to do everything *but* hide for dear
life. Let's first examine a subcategory of cross-dresser that isn't
sexually motivated, or at least for whom lust isn't the primary
inspiration. There are male "drag queens" who impersonate
women for their livelihood, for instance, but they don't necessarily dress this way for any sexual reason. (This is also true for the
less frequently seen "drag kings," female entertainers whose acts
are male impersonations.) It's usually money, a love of their craft,
or the thrill of performing (often all three factors, to different
degrees) that drives these people to gender bend, not their libidos. Then there are the male "transvestites," whose cross-dressing

habits are most definitely libidinal.* For transvestites, the primary turn-on is the feel of female garments (usually undergarments, such as a pair of matching panties and bra from Victoria's
Secret worn discreetly under a Brooks Brothers suit) as they rub
against their skin; lustful thoughts of women are elicited by the
texture, tactility, and other sensual attributes of the clothing. It's
not as common as sadomasochism, but neither is transvestism
rare: around 3 percent of straight men report having become
sexually aroused at least once in their lives by cross-dressing.†
The "straight" part is a central point here, too; there's no such
thing as a homosexual transvestite since, rather obviously, gay
men certainly aren't going to get turned on by wearing sexy lin-
gerie that makes them think of fornicating with a woman.

Yet another subcategory of people who wear the clothes of
the opposite sex are those in the transgender community. These
are individuals who describe having gender dysphoria, which is
the unpleasant sense of their biological sex (or chromosomal sex,
as in XX or XY) being out of sync with the subjective feeling of
their gender, a potentially very painful psychological experience.‡
The clichéd expression for gender dysphoria is "a woman trapped
in a man's body," but—as in the case of Cher's son, formerly her
daughter, Chaz Bono—transsexuality can also take the form of a

*As with the paraphilias more generally, there are virtually no female transvestites, or
women whose most intense sexual desire centers on wearing men's undergarments.
†A Swedish study found that transvestic fetishists are more likely to be sadomasochists
than the general male population. See Niklas Långström and J. Kenneth Zucker,
"Transvestic Fetishism in the General Population," *Journal of Sex and Marital Therapy*
31, no. 2 (2005): 87–95. When the "Voice of Basketball," the legendary sports broad-
caster Marv Albert, was arrested in 1997 on charges of rape and sodomy (he was alleged also to have violently bitten a woman on her back), one of his accusers added that
he was wearing women's panties and a garter belt when the assault occurred. See
Brooke A. Masters, "Albert Apology May Clear Record," *Washington Post*, October 25,
1997, www.washingtonpost.com/wp-srv/local/longterm/albert/albert.htm.
‡There are also several chromosomal perturbations of biological hermaphroditism,
which can complicate both the subjective sense of gender and the individual's expression of gender identity. But since our focus here is on the psychology of sexual deviance
rather than gender and biological sex per se, we won't be exploring these in any detail.

"man trapped in a woman's body." There's one big difference between male-to-female (MTF) transsexuals and female-to-male
(FTM) transsexuals, however, and this is the fact that whereas
the vast majority (around 75 percent in the West) of the former
are "heterosexual," nearly all of the latter are "homosexual."*

The language here gets dicey, to say the least. But it's not quite
as complicated as it sounds. Think of each of us as being com-
prised of three basic parts. First, we each have a biological sex,
which (except in rare cases of chromosomal disorders) is either
"male" or "female." This is what gets written down on our birth
certificates. Second, we each have a gender. Again, gender is our
subjective feeling of being male or female. Our gender usually
matches our biological sex but, as we've already seen, this isn't
always the case. Finally, each of us has a sexual orientation, which
means that we're erotically attracted to males, females, or—in
the case of bisexuals—both males and females. (There are also
asexuals, who have a lifelong pattern of being attracted to neither
males *nor* females.) The important thing to understand about
these three elements (biological sex, gender, and sexual orientation) is that for any given person they can combine in any number
of ways. Most people get the standard concoction, whereby whatever is jotted down on our birth certificates matches the feelings
in our heads of who we are, and we grow up to be most erotically
attracted to those of the opposite biological sex. But speaking as a
case of "male-male-male" on all three of these dimensions, which

---

*In Far Eastern countries such as Korea, Malaysia, Singapore, and Thailand that have
considerably less tolerance for effeminate men leading openly gay lives as males, it's the
polar opposite pattern. More than 95 percent of the MTF transsexual population in
those regions consists of biological males who are attracted exclusively to men. In their
early development as boys, these *kathoeys*, or "lady-boys," of Southeast Asia tend to be
extremely feminine in their mannerisms and appearance. They're not accepted as gay
men, but they do have an easier time than more masculine sorts in making "convincing"
(in other words, socially acceptable) women after their physical transition to female. See
Anne A. Lawrence, "Becoming What We Love: Autogynephilic Transsexualism Conceptualized as an Expression of Romantic Love," *Perspectives in Biology and Medicine*
50, no. 4 (2007): 506–20.

is to say, I was born a biological male, I've always felt like I was a male, and I've only ever been attracted to other males—deviations from the norm are not uncommon.

With transgender individuals, the labels we like to adopt for social identity reasons can look hazier, but beneath the terminological fog is the same basic three-factor combination logic. For example, because many MTF individuals are still attracted to women, they often adopt new identities as lesbians after undergoing surgical or hormonal changes to their physical appearance. Likewise, biological females who once identified as lesbians may come to see themselves as straight men after they've transitioned. But whereas the physical appearance of one's sex can change dramatically, for biological males at least, erotic tastes are pretty much a done deal once they're fixed in place. In other words, regardless of gender, if a biological male has a heterosexual orientation (that is, is "straight") for the thirty or forty years prior to transitioning to a female identity, after she becomes a woman physically, she's still going to have that same sexual orientation. (Just ask any wife—or ex-wife—who married a man who then became a woman. At no point was her spouse ever a "gay man.") The individual may identify as a lesbian now, but her sexual orientation is the same as before.

Here's where that considerable conflict I spoke of earlier rears its ugly head (and really, it's all gotten quite brutal, complete with harassment and social-media wars between the two opposing theoretical camps).* Whereas it's clear enough to most researchers that *homosexual* transsexuals aren't erotically motivated to permanently transform themselves into women (or men, in the

---

*This escalated to peak intensity upon the 2003 publication of the psychologist J. Michael Bailey's book, *The Man Who Would Be Queen: The Science of Gender-Bending and Transsexualism* (Washington, D.C.: Joseph Henry Press). In the book, Bailey embraced the controversial autogynephilia theory to explain heterosexual MTF transsexuality, with many people in the trans community being exposed to these ideas for the first time.

case of FTM individuals) but simply want to rid themselves of the horrible gender dysphoria that has gnawed at them their entire lives (more often than not, these are individuals who've lived as very effeminate males or as very masculine females since their early childhoods), some prominent sexologists believe that it's a different story altogether for *heterosexual* MTF transsexuals (who tend not to have as many stereotypically "effeminate" characteristics as their homosexual MTF cohorts). Thus, although it's often misunderstood, the controversial theory that I'm about to describe applies only to one specific subcategory of transgender individuals: those born as biological males, who have a female gender, and who've only ever been attracted to females.

The controversy over the "real" motivations of these biological males who are attracted to women dates back to 1989, when the psychologist Ray Blanchard postulated the existence of a paraphilia involving "a male's propensity to be aroused by the thought of himself as a female." He called this "autogynephilia." To Blanchard and others, heterosexual MTF transsexuals want to become women not so much to relieve their gender dysphoria as to actually incarnate their erotic target. And *that*, as you might imagine, hasn't sat well at all with the transsexual community, many of whom feel they're being falsely accused of lying when proclaiming it really *is* about gender dysphoria for them, not about lusting after themselves like some weird "pervert."

But Blanchard didn't just pull his autogynephilia theory out of thin air. In 1986, he'd invited a group of (pre-op) heterosexual MTF transsexuals to his lab and asked them to fantasize about wearing women's clothes and applying makeup, or essentially to simulate in their minds going about their own normal routine for transforming into their female identity. The rub in the study was that their male genitals were hooked up to a penile plethysmograph during this mental exercise so Blanchard could measure their degree of sexual arousal while they thought these gender-bending thoughts. He reported finding "significant" erections even for those individuals who said their cross-dressing had nothing at all to do with sex but was solely about being "women in spirit." Blanchard has pointed out that these data don't imply these men were lying; instead, he reasons, perhaps they just hadn't admitted their sexual feelings to themselves.

In any event, if Blanchard is correct, then autogynephilia is basically a more pronounced form of transvestism; it's not the clothes alone that arouse such men but the entire character and essence of the woman they seek to bring to life. Given the profound level of prejudice they continue to face (imagine being constantly and unapologetically referred to by many in society as an "it" or a "thing"), it's easy to see why the concept of autogynephilia has rankled those in the transgender community, but in addition to the findings from Blanchard's study there's a good deal of supporting evidence to be found in old case reports with transvestites. Here's a blurb from 1940, for instance, with an older crossdresser describing his earliest erotic memories. (I found this anonymous tidbit while burrowing in Alfred Kinsey's own labyrinthine files at Indiana University.)* "My sex life manifested around the age of 12 or 13," he explains:

> And it was immediately associated with my desire to wear girls' clothing. I began to discover that when I would put on a dress, skirt or high-heeled shoes I would almost immediately have an erection. In later years I asked numerous transvestites if they had ever experienced such a

*In her memoir, *Mirror Image: The Odyssey of a Male-to-Female Transsexual* (New York: Holt, Rinehart, and Winston, 1978), the now-adult MTF Nancy Hunt describes her adolescent feelings as a boy this way: "I was feverishly interested in girls. I studied their hair, their clothes, their figures . . . brood[ing] about the differences between us. I seethed with envy while at the same time becoming sexually aroused—I wanted to possess them as I wanted to become them. In my night-time fantasies, as I masturbated or floated towards sleep, I combined the compulsions, dreaming of sex but with myself as the girl" (60). And in her rather deliberately titled book, *Men Trapped in Men's Bodies: Narratives of Autogynephilic Transsexualism* (New York: Springer, 2013), the self-described autogynephilic transsexual therapist Anne Lawrence provides many similar anonymous accounts of an underlying erotic motivation as shared by her heterosexual MTF patients.

reaction and the replies were almost universally affirmative. Whenever I [needed] sexual gratification all I had to do was put on some feminine clothing, think about a boy dressed as a girl, or look at pictures of female impersonators and the gratification was very quickly satisfied. It was an association with the female figure with whom I identified.

Blanchard's theory of autogynephilia is one of the most battle scarred in all of modern sex research. Despite the antagonism on both sides, no attempt has yet been made to replicate his original laboratory findings, so the theory remains highly contentious. But valid or not, the very idea of autogynephilia is about as benign a paraphilia as I can possibly think of. (Essentially, one is aroused by oneself as an idealized member of the opposite sex.) Whether their "real motives" are erotic or the result of gender dysphoria, the personal distress so often experienced by *any* transsexual is the result of living a life ensconced as a harmless minority among an intolerant majority.

Sometimes, however, no matter the societal conditions in which it occurs, the paraphilia is indeed the direct cause of its own landslide of personal distress. Take the twenty-six-year-old sneeze fetishist from London. It's easy to get this bungled in one's mind with the practice of "nasolingus" (which, as we all know, involves passionately sucking on the nose of one's partner for sexual gratification), but this young man—an actor, by the way—had altogether different nasal affections. I can't tell you what it's called, because his paraphilia is so uncommon that no sexologist ever bothered to baptize it in Greek. But the patient's most arousing stimulus was being in the presence of a good-looking man in the grip of a sneezing fit. "Much of his intense interest," explains the psychiatrist Michael King, "was associated with a wish that he could sneeze in such an attractive fashion himself." Neither the physician nor the patient could be sure how or why this peculiar turn-on started. Yet in line with the sexual-imprinting model, he thought it might have something to do with the fact that the man had been frequently stricken with allergies as a boy.

Now, compared with placing wiggling invertebrates around your anus, a sneezing paraphilia sounds almost *cute*. But this fellow's unmanageable lust for that which prompts the kindest of "Gesundheits!" was utterly ruining his life. A sudden, unpredictable explosion of nasal secretions lurked around every corner, and such surprise fusillades led to the most embarrassing orgasms. Also, whenever he wasn't onstage, the man tried desperately to make himself more attractive by feigning "handsome" successive sneezes. But this only made him appear constantly sick and, as is the predictably adaptive response to such germ-ridden displays, caused everyone to avoid him like the SARS virus. He was in a relationship once, but he'd gotten so jealous of his partner's sneezing style that the latter packed up his bags and left, citing irreconcilable differences. Bless that guy's heart for trying, though. If you've ever seen a person's face as it's halfway through a sneeze, it's really not a good look.

Subjectivity, personal distress, harm, deviancy, children, innocence, moral judgment, individual differences, sexual imprinting, religion, society, mental illness, you can see how it all gets quite complicated. The whole business is messier than a sneeze fetishist's handkerchief at the end of a long day. But to see just how messy it is, you've got to lean in as we've been doing here. It's not always pretty up close, but it's better than remaining at a distance, where these important issues can appear deceptively simple. The public debate plays out in an infinite regress of blame over who's responsible for those who fail to fit the standard erotic mold. This is variously ascribed to the people choosing to be the deviants they are, porn, the Devil (always a shoo-in), bad parents, poor role models, our sexually repressed culture, or the psychiatrists who keep needling sexual minorities by branding them mentally ill. It's a rabbit hole of endless (and usually endlessly bad) arguments. Morally, all that matters—and allow me to reiterate that because I feel it's quite important, *all* that matters—is whether a person's sexual deviancy is demonstrably harmful. If it's not, and we reject the person anyway, then we're not the good guys in this scenario; we're the bad guys.

SIX

A SUITABLE AGE

Wilde: He was about 16 when I knew him.

Prosecutor: Did you ever kiss him?

Wilde: Oh, dear no. He was a peculiarly plain boy. He was,
   unfortunately, extremely ugly.

— Testimony of Oscar Wilde (April 3, 1895)

Let's leave the science labs, the courts, and the psychiatrists' offices for a moment and head over to Las Vegas. Picture yourself walking into a towering casino—into its arcade of electric din and seizure-inducing shimmering lights, past its garish foyer drenched in gold and mirrors, through its aisles of vacant-eyed regulars affixed like barnacles to their stools—and finally coming to stand before an imposing new slot machine. But this is no ordinary slot machine, you quickly surmise. Rather, it's a *sex* slot machine. That may sound fun, but the stakes here are massive. First of all, you're playing for someone else, someone yet to be born. In fact, he's yet to even be conceived. So although the game isn't going to affect you directly, a person's whole life is in your hands with this game. What's at play is the sexual destiny of this future human being. Basically, if you win big with your spin, you'll be securing for this individual a sex life of tranquillity, love, family, and acceptance. Lose big on the sex slot machine, however, and this person will have to face a lifetime of society's never-ending wrath and all that comes with it, including feelings of anxiety, self-hatred, shame, loneliness, and rejection.

Since you won't personally benefit by winning, it's free to play. And unlike with most slot machines, the odds are actually in your favor (or rather, in the favor of this unnamed soul you'd be playing for). The catch is you only get one go at it, and you're feeling pretty squeamish about this strange responsibility. On the other hand, for the child to ever come into this world someone has to pull the lever, and if it's not you, then the next person who comes along will do it. So you take a deep breath, picture in your head the innocent face of the child on whose behalf you're playing this devilish game, and reach out to give this one-armed bandit with the mechanical motives a good firm handshake . . .

Here's what you know. Like every other slot machine, this one has its own preprogrammed algorithm for randomizing the results. The only difference is that this algorithm involves a mix of factors that, together, will determine the individual's sexuality: genes, prenatal experiences, brain chemistry, early childhood events, family dynamics, cultural milieu, and an untold number of other inscrutably interacting variables. It's a complex system that's impossible to pick apart, let alone gain any control over. Cross your fingers or say your prayers for this kid, but no matter what superstitious rituals you perform while the wheels are spinning, he or she gets what he or she gets.

The rules of the game are clear enough at least. You can see, for instance, that the sex slot machine has four separate windows in its main horizontal row. Whatever ends up at each of these windows will reflect a distinct slice of the individual's sexuality. Behind the first window is the spinning wheel of *sexual orientation*. There are four possibilities here: heterosexual, homosexual, bisexual, asexual. The second window will reveal the person's ultimate *erotic target*, with the possibilities being person, animal, inanimate object, or none (if lined up with asexual). Behind the third window lies the dominant *erotic behavior*, with four possible outcomes: normal intercourse, courtship paraphilia (such as exhibitionism, voyeurism, frotteurism), other paraphilia (one of those infinite possibilities), and masturbation only. Finally, in the last window, you'll learn what *erotic age orientation* is in store for this person, which will be the focus of the present chapter. The spinning wheel behind this spot can land on one of six general outcomes: pedophilia (prepubescent), hebephilia (pubescent), ephebophilia (older adolescent), teleiophilia (mature adult), gerontophilia (the elderly), or none (again, if the person will be asexual). As with all slot machines, it's how the symbols line up in their totality that makes all the difference. It's the unique mixing of these traits, these four slices combined, that will constitute the erotic profile of this sensitive human-to-be.

Now, chances are you'll pull the lever and see staring back at you the arrangement of symbols occurring with the highest statistical frequency: heterosexual/person/normal intercourse/teleiophilia. If so, then congratulations are in order, since you've just won for this new member of our species today's social jackpot. You've secured for him or her the reality of a straight person whose biggest turn-on is to have normal sex with other consenting adults. (It's also "Ladies' Night" at the casino, by the way. If your soul-in-waiting is female, the odds have just skyrocketed in her favor; let's say it's now "99 to 1" that you've got this or one of the other winning patterns in the bag for her.)

But what makes the game so nerve-racking is that you never know what you're gonna get, as Forrest Gump would say. And in this case, pulling the lever on the sex slot machine is like reaching into a box of chocolates laced with all kinds of different poisons, each of which would cause its own distinctive and unpleasant social effect for your young charge. Given the weighted variables, the likelihood of you ending up with, say, bisexual/animal/courtship paraphilia/gerontophilia (which, I regret to inform you, could very well mean that your person will be into flashing his genitalia to male and female water buffalo on their last legs) is slim to none. Nevertheless, that odd pattern is within the realm of possibilities. More probable would be a one- or two-factor deviation from the most common alignment. Some of these wouldn't be so bad for your person's social calendar these days. If you wind up with heterosexual/person/other paraphilia (sexual masochist)/ teleiophilia, for instance, E. L. James and her *Fifty Shades* trilogy have helped take the sting of shame from a desire to be whipped or chained, and there will certainly be no shortage of sadists willing to oblige. Things could go a whole lot worse, though. After all, at the other end of the row is a doomsday one-deviation situation such as heterosexual/person/normal intercourse/pedophilia. And if the wheels of your person's sexual fate end up looking something like that, then he's in for a hard-knock life indeed.

The future for your fragile constituent is likely to be so difficult because what appears behind this last window is the stuff of society's worst nightmares, with the dreaded pedophiles being especially loathed and feared. It hasn't always been so. As we saw in the previous chapter with the flux in age-of-consent laws, our own "pedophilia panic," as some scholars have taken to calling it, is actually of fairly recent vintage. One revealing study comes from the sociologists Melanie-Angela Neuilly and Kristen Zgoba, who investigated historical trends in U.S. and French media coverage of child sex abuse. Performing a fifteen-year content analysis on articles appearing in the relatively liberal *New York Times* and *La Monde*, Neuilly and Zgoba found that the terms "pedophilia" and "pedophile" hardly occurred in either source until about 1995. The researchers trace the onset of our modern pedophilia panic to the domestication of the Internet in the mid- to late-1990s and the pestilence of cybercrimes involving children. Yet it was the Catholic Church scandal, they argue, that launched a meteoric rise of these anxieties in 2002. It's not just the word "pedophile" that's taken on a life of its own, either. The pair also found that the phrase "sexual predator" increased by *900 percent* between 1990 and 2005. What's striking about this newspaper study is that reported child sex-abuse rates in both countries were noticeably declining just as the pedophilia panic was dramatically rising.

I know it's a scary matter, but you've got to admit that operating with the default assumption that every stranger is a crafty pedophile with one wolfish eye secretly roving for children isn't exactly the sign of a healthy society. I'm certain that I'd be overprotective and paranoid about pedophiles if I were a parent (at least a parent to children that don't bark, chew on rawhides, and have tails—sadistic zoophiles *do* both frighten and enrage me). But irrespective of our own emotional investments concerning such hard-to-even-look-at topics, there's still an unbiased scientific reality out there, and we really ought to look at this reality for the sake of everyone's sanity.

Now, assuming you're not already a pedophile or a gerontophile, I suspect you'll agree by your complete inability to be turned on by an eight-year-old or an eighty-year-old that your erotic age orientation is no more a conscious choice that you've made than whether you're gay, bi, or straight. Your mom's uterus wasn't exactly like a Las Vegas casino (although I suppose that depends largely on what she ate while you were gestating), but nonetheless your own personal slot wheels started spinning the moment sperm met egg. And if you're male, they slowed to a full stop right around the time that you were busy blowing out the candles at your tenth birthday party. That, ladies and gentlemen, is the harsh reality of what we're dealing with here: just as a child can be said to be gay, a child can also be a fetishist, a sadist, and even, oddly enough, a pedophile. The hormones that will vivify a child's sexuality and make him or her (potentially) dangerous are a few years away still, but come what may, the slot wheels have more or less already come to a grinding halt.

Once he's an adult, a pedophile may completely agree with us and understand, intellectually, that only other adults possess the emotional maturity needed for a proper sexual relationship. That's all well and good, but since adults don't do anything for him sexually, it's also entirely useless. Telling a pedophile that he needs to be attracted to grown-ups, not kids, is like telling a lesbian that she just hasn't found the right guy, trying to convince a transsexual woman that her gender dysphoria is only a phase, or attempting to persuade a straight man that he'll really enjoy being anally penetrated by another man (maybe you've had better luck with that last one than I have). So if you really want to eradicate child sex abuse, you're not going to do it by just standing there with a giant judicial broom in your hands and sweeping the endless flow of adult pedophiles into prisons and psychiatric facilities as they continue to spill out of the ether. You need to first figure out why there are so many of them and how, exactly, they've come to be this way.

But doing so is far from simple. A good first step is acknowledging that pedophilia is indeed a sexual orientation. And like any other sexual orientation, its causal antecedents lie in early development, not in the adult's perverted, against-what-is-right choice to "become" a pedophile. Even in science, we're not quite to the point of being able to take such an objectively amoral approach to this issue, but once we are, research on other, less worrisome, sexual orientations may offer some directions for how to go about studying the childhood traits that best predict if a child will grow up to be a pedophile.

In the early 1990s, the sexologists J. Michael Bailey and Kenneth Zucker started using the term "pre-homosexual" to refer to young children who were likely to grow up to be exclusively attracted to the same sex. Drawing from a variety of methods (including long-term studies that followed children as they grew up as well as retrospective studies in which gay and straight adults reflected back on their own childhoods), the researchers pinpointed a set of factors that could reliably forecast adult homo- or heterosexuality beyond statistical chance. Basically, the old stereotypes proved to have the best predictive value: there were plenty of exceptions, but generally speaking, displaying "gender-atypical" traits in early childhood was most closely associated with adult homosexuality. The little girls who gravitated to rough-and-tumble play with boys were more likely to become lesbians, for example, whereas the little boys who oriented to dolls and dress up with girl playmates were more likely to become gay men. We may not like the fact that these old stereotypes have any basis in reality, but irrespective of that, they do.* What Bailey and Zucker report is only a statistically meaningful trend, of course, not some inviolate law of the queer universe. There are plenty of traditionally masculine boys who grow up to be gay as well as girly future lesbians who love dolls. But the authors did find a strong "dosage" effect, so that the greater the number of gender-atypical traits found in a given child, the more likely he or she was to grow up to be gay. (I, for one, was a bit of an androgynous mix on these measures; I preferred hopscotch to football during grade-school recess, but when I did play sports, I was known as a ruthless little hack with a passion for kicking shins.)

In principle, a similar predictive approach could be used to determine which children are most likely to grow up to become "minor-attracted adults" (as many pedophiles and hebephiles have begun referring to themselves in an effort to distance themselves from those less appealing descriptors). Obviously, calling

---

*"A man crosses his legs by resting an ankle on his knee; a sissy drapes one leg over the other," wrote the novelist Edmund White in A Boy's Own Story (1982; New York: Penguin Books, 2009). "A man never gushes; men are either silent or loud. I didn't know how to swear: I always said the final g in *fucking* and I didn't know where in the sentence to place the *damn* or *hell*" (8).

their seven-year-old a "pre-pedophile" isn't going to go over well with most parents, but that doesn't mean that the underlying construct isn't valid. And if we're sincerely motivated to reduce harm to children, an empirical endeavor such as this can go a long way; we might not be able to change or "cure" their erotic age orientation by the time we're able to scientifically identify them, but if the matter is handled sensitively, we can certainly intervene by educating these young people about their unique burden and the responsibilities that come with being who they are. This type of rational response can also quell these pre-pedophilic children's fears of rejection. Their dawning awareness that they're getting older but remain attracted only to those who are much younger will inevitably fester into internalized symptoms of personal distress in the absence of any social support. (One recent study found that pedophiles and hebephiles begin recognizing—and subsequently worrying about—their taboo sexual natures around the age of sixteen or seventeen.) That's not good for any of us. Having disgruntled, misanthropic adult pedophiles walking around and blaming society for their cruel lot in life is just lighting a fuse; this will only encourage them to self-justify and rationalize any harm they'd cause, putting more children at risk.

Yet aside from being logical and reasonable, elements that tend not to survive very long in the social atmosphere of a moral panic, such a preemptive social approach, in order to work, also requires a much better scientific understanding of pedophilia than what we currently have. After all, unlike many pre-homosexual children who stand out even to the casual observer as being likely to grow up to be gay, prospective pedophiles don't have such blatant stereotypes for scientists to work with in studying them. There's no obvious pre-pedophile equivalent of a "sissy," that's to say, and given that the only openly pedophile adults you'll probably be able to find have been ignominiously outed by the authorities after committing some crime, these people's retrospective childhood accounts may not reveal the traits of a pre-pedophile so much as a history of impulse-control problems.

Complicating matters even more, there are two distinct sub-types of pedophile offenders in the "system," according to the forensic psychiatrist Michael Seto. There are those whose crimes were hands-off (typically, child porn offenders), and then there are those who've committed hands-on offenses (pedophilic child molesters). These aren't mutually exclusive categories, clearly, but they're just as clearly not one and the same. So let's first examine what the science tells us about the hands-off variety.

Seto reports that only one in eight men arrested for child porn possession is known to have committed a hands-on offense against a child and that, despite the popular view that men who view child porn will eventually harm an actual child because the visual fantasy alone will eventually not be enough, the most compelling statistics actually speak to the contrary. Overall, Seto argues, these individuals are less likely to display the empathy deficits more characteristic of a hands-on offender. Although they're pedophiles, they're not necessarily cruel, and their normal supply of empathy helps to prevent them from physically acting out their desires with a real child.

One of the more controversial studies on the rightfully volatile subject of child porn is a 2011 report by the biologist Milton Diamond. Similar to previous work showing that the more sexually explicit materials in a society, the fewer reported sex crimes against women (presumably with porn functioning as a sort of masturbatory "displacement" for nonconsensual sex in potential male offenders), Diamond showed that societies in which child porn was at one time legal had lower reported rates of child sex abuse during those periods. Diamond arrived at this startling conclusion by analyzing crime statistics in the Czech Republic. From 1948 to 1989, the country's Communist regime strictly prohibited all forms of sexual expression (even the relatively mild *Playboy* and cheap romance novels were forbidden, although I'd probably still have been able to get my beloved muscle mags, mercifully). By late 1989, the regime had fallen, and porn quickly became a booming cottage industry in the newly democratic region. Almost overnight, Czechs went from absolutist laws barring the use of sexually explicit material to a completely unregulated marketplace where any type of porn, including child porn, was easily and legally obtained. In comparing the rates of child sex abuse in the seventeen years *before* the revolution with those during the eighteen years (1989–2007) *after* the revolution, Diamond uncovered a precipitous drop in reported child sex-abuse cases as well as sex crimes against women. (Nonsexual crime rates, by contrast, increased during the same postrevolutionary period, a fact that makes the sex-crime data difficult to explain away as the result of some rising general happiness quotient linked to these sweeping new political changes.)

Diamond's data speak for themselves, really, so there's in fact little reason to doubt them, nor his theorizing about the sexual catharsis effect of giving porn to the masses. But it's the nature of this particular porn category—involving *kids*—that makes it so difficult to accept. Nonetheless, researchers analyzing sex-abuse statistics in Japan and Denmark have also found that the legal availability of child porn is associated with reduced overall rates of child molestation (and in fact many pedophiles and hebephiles who use child porn will tell you the same thing, that these materials help to keep them from harming children). Yet it's also difficult to know what to do with these data in practice. I mean, sure, fewer children may be molested due to child porn availability, but what about those children being abused and exploited in the production of this "displacement" material? It's not displacement for them; it's their actual, flesh-and-blood exploitation and abuse. This isn't an airy hypothetical problem one tosses out in philosophy seminars as a sort of morally gray ethical dilemma; it's a very real scenario involving very real children. Diamond and his Czech coauthors recognize this not insignificant problem as well. "We do not approve of the use of real children in the production or distribution of child pornography," they write, "but artificially produced materials might serve." In most places in the world, however, computer-generated child porn is strictly illegal too, so that's not presently a viable option. It may make us personally squirm, but banning synthetic children for pedophiles' private pleasures actually makes little sense when considered within the moral framework of harmfulness that we've been using throughout this book. Diamond's data strongly suggest this ban ultimately will add up to more *real* children being harmed. In the absence of mindless (and therefore impossible to harm) erotic targets such as synthetic children, the only other alternative that manages to strike both a measure of pragmatism and child protection is a sort of government-controlled allocation of confiscated child porn to diagnosed pedophiles, particularly those deemed by clinicians likely to offend (or reoffend) by committing hands-on crimes. Perhaps only dated illicit material that has been approved for this use by the now adult models used in the images would be employed in this way. It certainly isn't a happy scenario, but it could be a practical one where a tone of rationalism is sorely needed. With the modern Greek government already recognizing pedophilia as a mental disability deserving of state-sponsored pay benefits, the "medical" procurement of child porn in an effort to reduce harm to children may not be so far away. As the philosopher Michel Foucault once said, "When the monster violates the law by its very existence, it triggers the response of something quite different from the law itself. It provokes either violence, the will for pure and simple suppression, or medical care and pity."

Meanwhile, a child porn conviction is a much better indicator of pedophilia than a child molestation conviction. The phrase "nonpedophilic child molester" sounds like an oxymoron, but in fact it's not, since a large proportion of child molesters (shown in studies to be at *least* half) abuse children as sexual surrogates for adults. These so-called opportunistic offenders are often inebriated, sadistic, or otherwise afflicted by some short-circuiting error in their frontal lobes. From a clinical point of view, they aren't "true pedophiles," because in the lab they demonstrate a greater sexual response to nude images of adults than to kids. It may be helpful, conceptually, to think of these opportunistic offenders as being similar to Kinsey's farm-bred males who couldn't find a woman to have sex with, so instead they had sex with a goat (or some other nonhuman mammal grazing in the pastures). Just as the latter weren't "true zoophiles," neither are the former truly pedophilic. Rather, they simply violate whatever, or *whoever*, is available to them at that very unfortunate moment and use their imaginations.

By contrast, those recurrently seeking out porn that features prepubescent children (or pubescent children in the case of hebephiles), when there's a bottomless well of more-than-available grown-up porn, are obviously exhibiting an orientation to minors. In any event, while these two subtypes of criminal pedophiles—hands-on and hands-off—aren't always distinguished from each other in studies, we do know a few of the traits that are associated especially with the hands-on variety. Before I share these with you, just be aware that these are statistical correlations only, which means that the co-occurring relationship between pedophilia and these traits is greater than would be expected by chance alone. Even if someone you know has all of these attributes (such as, ahem, that new boyfriend of yours), it's still exceedingly unlikely that he's attracted to prepubescent children. (Likewise, just because someone doesn't have any of these traits doesn't mean he's *not* a pedophile.)

The psychologist James Cantor, whom we met earlier while distinguishing the erotic-target paraphilias (who or what you're attracted to) from the erotic-activity paraphilias (what you like doing), has uncovered a number of these patterns. For reasons presently unknown, for example, pedophiles are more likely than other men to be short in stature. Another finding is that there is a disproportionate number of left-handed pedophiles. Being a southpaw is usually just a genetic quirk that means nothing in particular, but it can also emerge as the result of damage to the cerebrum during prenatal development. So the overrepresentation of left-handedness in pedophiles hints at a neurological "born that way" basis of having a sexual interest in children (or perhaps, born with a predisposition for getting sexually imprinted that way). Cantor uncovered another revealing neurological trend among pedophiles who sexually abuse children: the younger their victims, the lower the offender's IQ. It's by no means a perfect correlation between IQ and erotic age orientation (if it were, all gerontophiles would be geniuses and all pedophiles would be drooling idiots, neither of which is true), but it does say *something* is different in the actual brains of pedophilic child molesters.

Cantor's big scoop, however, came when he discovered anomalies in the density of white matter in pedophile neuroanatomy. "The 'white matter' is the shorthand term for groupings of myelinated axons and glial cells that transmit signals throughout the gray matter that composes the cerebrum," he explained to a reporter in 2012. "Think of the gray matter like the houses on a specific electricity grid and the white matter like the cabling connecting those houses to the grid . . . There's either not enough of this cabling [in pedophiles], not the correct kind of cabling, or it's wiring the wrong areas together, so instead of the brain evoking protective or parental instincts when these people see children, it's instead evoking sexual instincts. There's almost literally a crossed wiring."

Yet even if it's all a matter of white matter, this still tells us little about how the matter came to be. That's to say, we know next to nothing about how to prevent mothers from giving birth to pre-pedophilic babies in whose soft heads the bungled white matter cabling process is already well under way. We run into the same problem as we did when trying to come up with (ethical) experiments using children that can isolate the specific cause of any paraphilia. In fact, it's unthinkable in the present case. The ethical problems of turning kids into bee-loving melissaphiles for the sake of science pale in comparison to the atrocity of randomly assigning newborns to be in the "pedophile" or "nonpedophile" condition of a controlled experimental study. Imagine being the parent of a child in the pedophile condition and waiting patiently to see if the experimenter's hypothesis pans out, with your bouncing baby boy (or girl) growing up to be an adult sexually aroused by six- and seven-year-olds.

As an example of how difficult it is to infer the causal origins of pedophilia when you're dealing only with correlations, consider the findings from a questionnaire given to thirteen hundred anonymous men in Finland. Men who reported having childhood sexual interactions (such as "playing doctor") with other children were now more likely, as adults, to express a sexual interest in children under fifteen. On the surface, this might *look* like evidence of male sexual imprinting. And it may well be. But there's another way to interpret this correlation that's just as plausible. It could be that these men weren't imprinted by these early experiences; instead, they were pre-pedophilic boys who'd been more motivated to engage in sex play with other children.

There's also no clear evidence to support the common assumption that being molested by a pedophile causes victims to become pedophiles themselves. There *are* data showing that being molested leads to a greater likelihood of sexually abusing a child oneself (the disturbing "abused abuser" cycle effect, which is usually explained by some neo-Freudian power-and-control theory or model of social imitation), but this doesn't necessarily have anything to do with the victim turned perpetrator's own erotic age orientation. If anything, if the abuse occurs at a sensitive period of male development and leads to the child's incidental arousal, he'd be more likely to develop a paraphilic attraction to harmful adults like the offender or to erotic cues that resemble the abusive event—an unsettling imprint in its own right.

~

Similar to the huge gap in the sex ratio for all the other paraphilias, there are very few certifiable female pedophiles. Some sexologists aren't convinced they exist at all. You're probably shaking your head in disbelief on hearing that, given the abundance of stories in the media about predatory "pedophile" female teachers. But remember that a pedophile is someone whose primary attraction is to prepubescent children. Although the word is often used to loathingly describe any adult who has sex with someone under the age of eighteen, a female teacher caught in a sex scandal with her hairy-chested high school student is no more a pedophile than a twenty-year-old having sex with a forty-year-old is a gerontophile. That's to say, unless they're, respectively, very late bloomers or have been smoking a whole lot of meth, most older teens aren't prepubescent and most forty-year-olds don't look anything like an elderly person.

Yet although they may be incredibly rare, case studies indicate that a handful of "true pedophiles" do exist among women. There are very upsetting incidents in which women have sexually abused shockingly young children. And although the act of child molestation itself isn't a direct indicator of pedophilia, targeting young children, combined with recurring fantasies limited to sex with prepubescents, is indeed so. In his book *Pedophilia and Sexual Offending Against Children*, Michael Seto describes several such women. In one case, a young mother reported herself to child protective services after she performed fellatio on her one- and three-year-old sons and found herself

masturbating to the memory of the incident.* Another woman
performed oral sex on a pair of four-year-old girls in her care and
became aroused while bathing them; her sexual fantasies were
similarly confined to young children. By and large, however, the
data on this mercifully slim demographic indicate that the majority of female child molesters aren't pedophiles but usually
are timid women who've been coerced by pedophilic men into
joining them in committing their crimes. More often than not,
these women are also the victims of abuse (in all its various
forms) and have a mortal fear of adult males. Consider also
that women represent less than 1 percent of all child porn offenders, an astonishingly (and tellingly) low percentage given
that women are thought to consume around a third of all porn
generally.

To scientifically confirm that a particular woman is a pedophile, researchers *could* use vaginal vasocongestion to measure
her arousal to images of nude models of different ages, just as
they do for men using the penile plethysmograph (as we'll see
shortly). But this is a subject area driven by forensic investigations, and in case you haven't noticed, sex offenders are vastly,
overwhelmingly male. Nevertheless, a speculative hypothesis
that's been floating around for a while, and one stemming from
those female genital hyper-responsiveness data we reviewed earlier, is that women should be more likely than men overall to

---

*There are massive cultural differences in parents' attitudes to their children's private
parts; what would be seen as sexual abuse by most of us is normal in other parts of the
world. In the 1940s, an anthropologist studying the Sirionó Indians of Bolivia noted,
"Parents are very proud of a display of sexual desire on the part of their infants. One
afternoon," he recalled of an incident with a Sirionó father, "Eantándu was fingering
the penis of his young son who was sleeping. The boy got an erection. Eantándu called
my attention to it and proudly said: 'Very hard penis; when grown, he will have a lot of
intercourse.'" Like the Sirionó, the Hopi of North America were also known to stimulate the genitals of their young children as a means of soothing them. Grandmothers
would mouth the genitals of their grandchildren because they believed it pacified col-
icky babies. Alorese mothers of Indonesia also fondled their infants' genitals while
nursing.

display sexual arousal to children and the elderly. It's an empirical question, and I'm guessing it's also one that won't get answered anytime soon, since ladies won't exactly be lining up to help researchers answer it. Yet even if it were found to be true, the crime data clearly show that females are still about as likely to act on their subconscious and unwitting genital arousal by eight- and eighty-year-olds as they are to date a bonobo. Remember, Meredith Chivers's ready-for-anything "protection hypothesis" assumes that vaginal responses to any and all sex cues—even highly unappealing ones (ape sex, rape scenes, naked men if you're lesbian, naked women if you're not, and *possibly* kids and octogenarians too)—would have evolved to reduce physical injury.

Again, a woman's genitals can be in stark disagreement with her desires.

The erect penis, by contrast, is a direct window into a man's erotic soul, or even more to the point it's a divining rod to his reservoir of specific desires. We've already come across a few studies during our journey in which the penile plethysmograph has been put to use in evaluating various scholarly questions about human sexuality. Its more common use is a forensic one, with the courts ordering sex offenders to undergo testing to distinguish the "true pedophiles" from the opportunistic offenders.* But this erection-detection machine was originally invented

*It's not a perfect predictor, but knowing the erotic age orientation of a sex offender is useful to law enforcement because pedophiles who *do* abuse children are more likely to re-offend than are the opportunistic offenders. After all, the former have an erotic age orientation that isn't going to go away, and by offending, they've already demonstrated their difficulty in controlling their sexual urges. The latter, by contrast, can attribute their crimes to a more transient problem, or something other than having an exclusive sexual interest in children, anyway. (The perfect-storm equivalent of a child molester is a pedophile who clinical tests reveal is also a certifiable sociopath. Very few pedophiles actually fit this bill, but such a person is an enormous threat to children.) See Seto, *Pedophilia and Sexual Offending Against Children*.

for another reason altogether. In fact, the scientific measurement of tumescence to ascertain a man's erotic tastes (a penis lie detector, of sorts) got its start in the most unlikely of places: the Czechoslovakian army.

In the 1950s, the Czech military approached the psychiatrist Kurt Freund, a practicing physician with a clinic in Prague, for help with a problem. It seems that a growing number of army recruits had been fibbing about being homosexual to avoid compulsory service, since openly gay men didn't have to enlist. So the top brass hired Freund to develop a reliable method of separating the gay wheat from the camouflaged straight chaff. They had come to the right man for the job. Like the army officials, Freund distrusted self-reports of people's sex lives. He was also more interested in the palpable biology of desire than he was in those fuzzy psychodynamics devised by a similarly named theorist from Vienna. Instead, what Freund needed here was a concrete measure of sexual orientation. And he quickly realized that a single dumb erection in a man would be more useful for the army's specific needs than an hour of prevaricating conversations with him in the clinic. The penis, Freund observed, could forever render impotent the art of the verbal lie when it comes to secret desires. (Freund also knew a thing or two about creating ruses in order to survive, or at least how to keep a low profile. Not only had he lived through the Holocaust as a Jew caught in the thick of the Nazi occupation of his native country in the early 1940s, but he somehow managed to avoid being deported to the concentration camps altogether. His parents and younger brother weren't so fortunate, however. All three perished at Auschwitz.)

The specifics have gotten more complicated in the decades since Freund first hand-delivered his sparkling new penile plethysmograph machine to the homophobic generals, but the basics of the procedure have remained largely the same: A man sits down in a chair, his penis is connected to an erection gauge that can pick up very subtle changes in tumescence (the device is so sensitive that it can detect a blood-volume increase of less than one cubic centimeter, which most men wouldn't even experience consciously), and he's then shown randomized images of nude models representing distinct erotic categories.* The scientist, meanwhile, measures what's happening with the man's own equipment as these photographs appear. In the first instance, the nudes were chosen to expose the fake gays in the army's recruiting pool, so they were simply static images of attractive men or women. But once the device had effectively ensured that any soldiers whose organs saluted vaginas were in place guarding his nation's borders, Freund began to see other applications for his new invention, including using plethysmography to identify closeted pedophiles. After all, if a straight man were asked outright if he finds some pretty seventh grader sexually arousing, he might say something on the order of "Oh, her? Well, she looks like a nice girl, but I wouldn't say she's 'attractive.' I mean, she's a bit young, isn't she?" What Freund wanted to know was the extent to which his penis concurred with that socially appropriate verbal sentiment.

Nowadays you can get similar results from looking at a man's computer hard drive, which police often do in the case of child porn investigations. But Freund's device is still routinely used to confirm (or rather to "diagnose") the defendant's erotic age orientation. Today, your plethysmography experience in a lab in Toronto, for example (where Freund relocated in 1968 after being targeted by his country's Communist Party as an intellectual dissident), would probably look something like the "Walking

---

*A small inflatable cuff is placed around the base of the phallus. The cylinder is then tightly secured over the cuffed organ so that the air inside is isolated from the atmosphere outside. A rubber tube attached to the cylinder connects to a pressure transducer. This, in turn, converts air pressure changes affected by increases in penile volume to some measurable voltage output for the scientist to read. See Kurt Freund, "A Laboratory Method for Diagnosing Predominance of Homo- or Hetero-Erotic Interest in the Male," *Behaviour Research and Therapy* 1, no. 1 (1963): 85–93.

Nudes" version of the test originally devised by Freund but per-
fected by Ray Blanchard. That name may ring a bell, since it was
Blanchard who came up with the controversial autogynephilia
theory of MTF transsexuals. As a young psychologist then working in the Canadian prison system, Blanchard also became one
of Freund's most trusted North American collaborators. It was a
relationship that would last until 1996, when, crippled by excruciating pain from an incurable lung cancer that had spread to the
lymph nodes in his neck and was now threatening his brain,
the inventor of the plethysmograph swallowed a fistful of muscle
relaxants and pain pills with a bottle of wine and fell dead to the
floor at the age of eighty-two. Yet despite the device's controversial use, it has lived on. And by the time Blanchard's Walking
Nudes Test (which I'm about to strap you into) appeared in 2002,
forensic investigators had anticipated almost every trick in the
book that a pedophile with his penis caught in a Freundian trap
might use to throw them off.*

The first part of the test, of course, involves the most essential part of all: your own telltale penis. Begrudging or not, under
court order, it will be expertly hooked up to the device by a
knowledgeable technician. The lights will then be dimmed as
you settle into a cozy reclining armchair. "Make yourself com-

---

*In 1976, the psychologists Vernon Quinsey and Sidney Bergersen discovered that
some pedophiles had been deceiving forensic technicians by the act of "pumping" on
the task, which is the voluntary contraction of the abdominal and perineal muscles to
momentarily "inflate" the flaccid penis and thus have it register as an erectile response
to the images of adults. (In the nonforensic world of bladder control and prenatal exercise classes, this is usually referred to as Kegel exercises.) Pedophilic or not, my male
readers can tend to this exercise easily enough in order to grasp how it would work; I'm
afraid there's no clear way to convey the exact process to my female readers except to
suggest that you ask some helpful male, preferably one you know, to demonstrate with
his limp organ. In addition to pumping, some pedophiles took to sneaking into the lab
a pin to surreptitiously prick themselves when images of children appeared, with the
pain killing their arousal. See Vernon L. Quinsey and Sidney G. Bergersen, "Instruc-
tional Control of Penile Circumference in Assessments of Sexual Preference," *Behavior
Therapy* 7, no. 4 (1976): 489–93.

fortable," you might be told sympathetically. You're even kindly given a blanket for your modesty. The more tranquil your surroundings, after all, the less anxiety there will be to interfere with your arousal to whatever it is you're about to see. (In the pioneer days of plethysmography, it was even fairly common to give the man some alcohol beforehand, just to loosen him up. "A certain degree of relaxation can possibly be achieved this way," explained the hospitable Freund.) If you're the extra-nervous sort and normally can't get it up under these types of conditions, or if you have some trouble experiencing erections due to your health or age (impotent old men can be child molesters too, don't forget), well I'm sure you'll do just fine, since it's become increasingly routine these days for technicians to give their male subjects a hefty dose of Viagra before getting started.

If you're worried about your penis's poor behavior—you've learned over the years that it's not always so obedient to what it *should* be doing—you might feel like closing your eyes and refusing to look. To pull your emergency erotic kill switch, maybe you can whip up some disgusting images in your head featuring genitals with STIs. Perhaps the right incestuous scene will do it, like performing oral sex on your orgasmically groaning grandmother . . . and toss in a nasty case of genital warts for her too while you're at it (it's not a pleasant thing to imagine nor, sadly, the thought of a loving grandson, but you know, it's a good last resort when you're in a locker room filled with hot naked straight guys). This metacognitive technique works the other way around also; in this case, if you're not attracted to grown-ups, you could picture in your mind's eye a child whenever a nude adult pops up on the screen. In Nabokov's *Lolita*, for instance, the hebephilic protagonist Humbert Humbert was able to consummate his marriage to Lolita's clingy thirty-year-old mother, Charlotte Haze, only by imagining that it was her twelve-year-old daughter ("light of my life, fire of my loins. My sin, my soul. Lo-lee-ta") that he was really having sex with. "Humbert was perfectly capable of intercourse with Eve," Nabokov wrote, "but it was Lilith he longed for."* But if *this* is your big strategy, it's not going to work in the Walking Nudes Test, since, oh yes, I forgot to mention there's a retinal camera trained on your pupils to make sure you're watching intently.†

And what you're going to be watching so intently on the three separate projector screens before you are films of three smiling models of the same approximate age and wearing only their birthday suits. Each will be advancing slowly, tantalizingly, directly toward you. That's to say, these visual stimulus sets comprise nude children and adults serving as erotic ambassadors, or perhaps "martyrs" is a better word here, for the pedophile, hebephile, and teleiophile categories.‡ As the models approach, the camera will zoom in on their genitals. All the while, just as it was in the days of Freund's original test, the forensic technician will be monitoring *your* genitals.

These aren't prurient images; the erotic triptychs are more anatomy-lesson film clips than porn. But not only does the Walking Nudes Test add movement and the illusion of the models' personal attention, it pads the image bank with multiple models from each age group rather than just a single representative. So, for example, if you're a hebephile, the odds of your seeing at least

---

*At least one scholar, Brandon Centerwall, has argued convincingly that Vladimir Nabokov himself, not just the fictional belletrist Humbert Humbert that the author famously created, was a closet hebephile (or "pedophile," as Centerwall calls him), with the book being Nabokov's attempt to exorcise a wanton demon haunting him all his life. See Brandon S. Centerwall, "Hiding in Plain Sight: Nabokov and Pedophilia," *Texas Studies in Literature and Language* 32, no. 3 (1990): 468–84.

†This ensures that you're not closing your eyes or looking offscreen as the nude models appear before you, but pupil dilation also correlates with sexual arousal.

‡Men's attraction to older adolescents is considered "normal" by most forensic psychiatrists—"barely legal" or "teen porn" isn't exactly a rare niche in the adult entertainment industry—so the ephebophile category usually isn't included. Instead, the teleiophile category may include young adult models in their late teens or early twenties. "True gerontophiles," on the other hand, are so rare that elderly models would only be included if the sex-offender subject has shown a history of targeting such-aged victims.

one of the pubescent nudes as a genuine "nymphet" or "faunlet" (Nabokov's term for the male version of a nymphet) are increased. Consider, then, that for the youngest ages, parents have volunteered their five- to ten-year-olds' gooseflesh to serve the purpose of weeding out the pedophiles from among the thousands of sex offenders (like you) taking part in this forensic study. Likewise, on any given trial you'll see the three sacrificial nude adolescents heading your way whose parents have agreed for their bodies to be used to detect men who grow tumescent for pubescents. Finally, of course, there are the naked adults perambulating for your viewing pleasure (we'll see about that, you sex offender, you), each of whom is a shining example of a fit, reproductively mature human being with a full suite of secondary sexual characteristics. You'll be exposed not only to models of your preferred gender, by the way, but both girls and boys, women and men, to sort out exactly what's behind that first slot machine window of yours, too. (There's some evidence that pedophiles are more likely to be bisexual, for instance, than are men of other erotic age orientations.)*

All of this is to say that your penis is about to pigeonhole you into a crucial category: Are you a pedophile, a hebephile, or a teleiophile? Bear in mind (and this may help some of you men breathe easier) that the Walking Nudes Test, and every other version of the phallometric method, is meant to determine your *primary* age of attraction, not necessarily your *exclusive* age of attraction. That's to say, it's not uncommon for a teleiophile to

---

*Incidentally, if you're a blind sex offender and can't see the nude models, don't feel excluded, because some early research has shown that an *audio* version of the standard plethysmograph may be just as effective. In Blanchard's "Narratives Slides Test," the images are paired with provocative audio narrations to amp up the male subject's arousal. For instance, you might hear through a set of headphones: "You are with your neighbors' 12-year-old daughter . . . you have your arm around her shoulders and your fingers brush against her chest. You realize that her breasts have begun to develop." See Ray Blanchard et al., "Pedophilia, Hebephilia, and the DSM-V," *Archives of Sexual Behavior* 38, no. 3 (2009): 339.

exhibit some tumescence in response to the images of pubescents, or even, occasionally, to those of the prepubescent children. This is why an adult teleiophile may recall being attracted to a fellow seventh grader at the time, though he wouldn't dare think of a twelve-year-old girl now in that way; even back then, he was still more attracted to his busty twenty-three-year-old science teacher than he was to the girl his own age, but he also wasn't entirely erotically immune to that cute twelve-year-old female, then *or* now. Nor is every pedophile or hebephile entirely phallically unresponsive to adults. The question is which age category is going to induce your strongest arousal response when your erectile averages are compared across the board. (This is generally the case with all the paraphilias; paraphilic men are often capable of some degree of arousal to erotic stimuli outside their particular kink, but the kink is undeniably their most trusty trigger.)

Most men, it's safe to say, know perfectly well what their erotic age orientation is before ever setting foot in the lab scenario I've just described. Yet that know-thyself reality can get a little confusing, given that children develop physically at such different rates. Some teenagers come with all the trappings of an adult—trappings colloquially known as "jailbait." It's been declining for years, but today the average age of menarche (a girl's first period) in the industrialized world is around twelve. But there's tremendous variance. Genes, socioeconomic status, stress, parental relations, diet, and a host of other complex factors are behind huge individual differences in female maturation, not only within the same society but in the very same suburb. Somewhere, a precocious fifth grader is shifting nervously in her chair after having to use a tampon in the school restroom while adjusting her bra; a more flat-chested girl, meanwhile, is pulling into the parking lot of the high school up the road, having recently gotten her driver's license but still waiting patiently for her first period.* Male development varies dramatically as well. I distinctly recall bragging to a friend in eighth grade about how I'd started sprouting a few odd hairs in my armpits, to which he replied in a voice deeper than my father's what a chore it had become to have to shave his beard every day.

Chronological age is of course *all* that matters in the legal sense. As we learned in the last chapter, it doesn't make the slightest bit of difference whether some chain-smoking, heavily tattooed fifteen-year-old with two kids looks more like a twenty-four-year-old escort; she's still clearly a minor in the eyes of the law. By contrast, for researchers studying the erotic age orientations, chronological age is almost entirely irrelevant. What's far more important to them is "biological age." These two constructs—how old one *is* (chronological age) versus how old one *looks* (biological age)—generally go hand in hand, but, again, it's not a perfect fit. A month of hormones working overtime can mean big physical changes to an adolescent's appearance. It could also mean that the man growing aroused by that particular adolescent isn't a hebephile but a teleiophile responding to the very adult physical cues being broadcast on the body of a child. In this area of forensic psychiatry, then, researchers want to know the age-related physical "type" that a man is most attracted to, not simply the specific chronological age of a person who happens to arouse him.[†]

To address this potential gap between chronological age and biological age, the models selected for use in today's

*Racial differences in age of menarche have also been used to discriminate against immigrant families, with lawmakers arguing that African American, Mexican, and Italian girls should have lower ages of consent than their white peers since they develop faster. See Stephen Robertson, "Age of Consent Laws," http://chnm.gmu.edu/cyh/teaching-modules/230.

†In parts of the Pacific, the average age of menarche is as high as eighteen. So technically, in some areas of New Guinea, a heterosexual hebephile could be defined as anyone aroused by seventeen- to twenty-one-year-old indigenous women.

plethysmograph studies are matched in terms of where they fall on the so-called Tanner scale. In his 1978 book, *Foetus into Man: Physical Growth from Conception to Maturity*, the eponymous British pediatrician James Tanner charted, in blushingly intimate detail that includes everything from the width of the areolae to the girth of the penis to the hue and texture of the vulva and scrotum, the precise physical changes that go along with each of six distinct stages of sexual development (from birth to full reproductive maturity) in both males and females. I can only imagine the auditioning process for becoming a Tanner stage model must be incredibly awkward for children and teens given such specificity of gonadal detail. And I say this having once had a Turkish pediatric endocrinologist who smelled of cigarettes and gauze pads thumbing my testicles with one hand while using the other hand to demonstrate to my parents how their thirteen-year-old son's dangling parts should be more walnut- than grape-size, so perhaps they ought to consider growth hormone injections (which I in fact received for several years). But the main point is that each of the Tanner stages is defined by a unique constellation of maturation-graded physical cues; whichever constellation of bodily traits best ignites your passions reveals your erotic age orientation.

At the time of that embarrassing doctor's visit, I'd probably have been in "Tanner Stage III," which on average is seen in boys around eleven to twelve and a half years of age. "Testicular volume between 6 and 20 ml," goes this clinical description, "scrotum enlarges further, penis begins to lengthen to about 6 cm, thicker pubic hair spreads to mons pubis, voice breaks, increase in muscle mass, may have some breast swelling (gynecomastia), sperm production may begin, growth accelerates to 7–8 cm per year." If the plethysmograph shows that particular type of organism to be the perfect one for you, then you're a homosexual hebephile . . . and quite possibly a priest.

Oh, I'm only joking about that last bit, of course. But as we've indeed seen from the Catholic Church sex-abuse scandals, in which 90 percent of the victims have been boys, homosexual hebephilia is one of today's most poisonous slot machine outcomes. You'd have hit the jackpot with this particular alignment back in ancient Greece, though; Plato famously claimed that pederasty was the one true feature distinguishing Hellenistic society from all the other "barbarians."* In fact, consider yourself lucky if you're a gay man who prefers men instead of boys. Throughout most of human history, you would have been rejected by society, or at least even more than you are today. (John Money, ever the rogue sexologist among his forensically minded peers, believed that some young boys are "androphilic," or aroused by adult men. "If I were to see the case of a boy aged 10 or 11 who's intensely erotically attracted toward a man in his twenties or thirties," Money said in a 1991 interview with *Paidika*, which was arguably less of a science journal than a pedophilia support group newsletter, "if the relationship is totally mutual, and the bonding is genuinely totally mutual . . . then I would not call it pathological in any way."†)

---

*The aristocrats of Athens, most of whom were men in their twenties and thirties, romantically courted boys of eleven or twelve as their special "pupils." A free boy was expected to allow an influential suitor to have exclusive intercourse with him (or at least "intercrural intercourse," sex between the thighs) in exchange for an elite education and sociopolitical benefits that would extend to the boy's family. In Xenophon's *Symposium*, the wealthy Callias bargains with the father of a boy named Autolycus for such an arranged relationship as the boy leans against his dad. What strikes the reader is the businesslike nature of the trade. There's no evidence that these boys were thrilled about this grown-up affair happening between their thighs, and some modern scholars believe that it was child abuse then just as it is now. See Enid Bloch, "Sex Between Men and Boys in Classical Greece: Was It Education for Citizenship or Child Abuse?," *Journal of Men's Studies* 9, no. 2 (2001): 183–204. (Actually, in *Phaedrus*, Socrates warns these young boys of the real motives of at least some of their idolized mentors: "Consider this, fair youth, and know that in the friendship of the lover there is no real kindness; he has an appetite and wants to feed upon you." See David West, *Reason and Sexuality in Western Thought* [New York: Polity, 2005], 23.) Others point out, however, that such institutionalized pederasty was widespread throughout ancient Greek civilization, and there are no obvious records of any mental or physical trauma suffered by these boys.
†*Paidika: The Journal of Paedophilia* ran from 1987 to 1995. Published by the Stichting Paidika Foundation, the mission statement of the journal read as follows:

In fact, two adult men of equivalent age and status being in a "versatile" romantic relationship (which is to say, they take turns being the insertive partner and the insertee) is a contemporary gay ideal. Basically—how do I put this delicately—in the past a man was often permitted his homosexual affairs as long as he wasn't the "bottom"; if he were the one doing the penetrating, his masculinity was unimpeachable. And this implied, needless to say, that a man could only have sex with males of lower social standing, which typically meant boys. This wasn't just the case in ancient Greece, either. Throughout ancient Asia, Australia, Melanesia, China, Japan, and most of the Islamic world, men are also frequently depicted in the historical literature as having intercourse with boys. A few centuries ago in traveling warrior societies such as the Japanese samurai and the Berbers of the Siwa oasis, pubescent boy "brides" accompanied jealous soldiers on their tours of duty while girls and women tended to things back at home. Today's ultraconservative Saudis might be interested to know that an eighteenth-century Englishman once observed how, on strolling by the Great Mosque at Mecca one day, he spied in its sacred halls men fornicating with adolescent boys while nonchalant passersby merely shook their heads and grinned. (Lesbians aren't without their share of hebephilic traditions either. For example, in the mid-1980s, a type of institutionalized lesbianism was discovered in Lesotho, in southern Africa. Prior to marriage, some young women in Lesotho courted pubescent

---

The starting point of *Paidika* is necessarily our consciousness of ourselves as pedophiles. To speak today of pedophilia, which we understand to be consensual intergenerational sexual relationships, is to speak of the politics of oppression. This is the milieu in which we are enmeshed, the fabric of our daily life and struggle. Through publication of scholarly studies, thoroughly documented and carefully reasoned, we intend to demonstrate that pedophilia has been, and remains, a legitimate and productive part of the totality of the human experience. (Stephanie J. Dallam, "Science or Propaganda? An Examination of Rind, Tromovitch, and Bauserman," *Journal of Child Sexual Abuse* 9, nos. 3–4 [2001]: 109–34.)

girls, the two then forming an intense romantic bond. These "mummy-baby" relationships, as they're called, provided not just opportunities for sexual experimentation but also emotional support for girls with unstable family lives.)

The homosexual factor aside, one of the most heated debates in this subarea centers on how hebephiles should be conceptualized and, more to the point, whether they should be treated as having a disorder. Most psychiatrists believe that pedophiles are mentally ill.* But there's a sharper division in the field over whether hebephiles are psychologically sick. That's to say, should those who are attracted primarily to pubescent children be added to the *DSM* as having a mental pathology, as pedophiles are conceived, or is a man's primary attraction to pubescents, although not as common (nor, certainly, as socially acceptable) as attraction to adults, still "natural" and "normal" enough to make such a diagnosis illogical? There are policy implications for whatever answer goes here. With the formal backing of the APA confirming that the individual is mentally ill due to his erotic age orientation, in some states a hebephile in prison for a sex offense (hands-on *or* hands-off) can be held indefinitely in a U.S. psychiatric hospital after he's served his sentence in full, if he's found to be at high risk of committing another offense in the future.

Making the determination of whether a sexual orientation is a genuine mental disorder has gotten increasingly challenging over the years. In 1992, the psychiatrist Jerome Wakefield suggested that for a trait to be considered diseased in this way, it must be biologically *dysfunctional*; that is, the trait must be at odds with an evolutionarily adaptive response. That sounds logical enough. But then if such a criterion were to be adopted by the APA or some other mental health organization as the defining

---

*Interestingly enough, not all psychiatrists consider even pedophilia a mental illness (Richard D. Green, "Is Pedophilia a Mental Disorder?" *Archives of Sexual Behavior* 31 no. 6 [2002]: 467–471.)

factor, one could reasonably conclude that homosexuality should never have been removed from the *DSM*, nor should other deviant psychosexual traits (such as "gender identity disorder," which could be used until 2013 to diagnose well-adjusted transsexuals with a mental illness) should ever have been plucked out.

When it comes to homosexuality, there are plenty of unconfirmed and circuitous theories about the possible evolutionary "reasons" for it (such as helping to raise your nieces or nephews, who share a quarter of your genes—and I'm afraid I'm a very bad uncle in this sense, living hundreds of miles away). But whether there's any truth to these Darwinian hypotheses or not, it's rather silly to argue over the blatantly obvious fact that being attracted only to the opposite sex is a much more effective gene-reproducing strategy than is being attracted only to the same sex. The APA's reclassifying homosexuality as a "normal form of human sexuality" in 1973 set an important precedent, because thereafter the psychiatric use of the word "normal" as applied to sexuality could never again be perfectly synonymous with "biologically adaptive."

Now, I realize how such a concession might encourage more than a few social conservatives out there to celebrate. "You *see!*" I can picture them saying. "I told you it was all just politics. The APA just caved to threats by the queers back in the 1970s, but gay people really *are* mentally ill!" Before such individuals break out the free champagne bottles that came with their membership in the Family Research Council, they might want to reconsider using "normal" and "abnormal" for the basis of their moral reasoning. After all, under the guidelines of reproduction and biological dysfunction suggested by Wakefield, the people that most conservatives would erroneously refer to as "pedophiles" are, in fact, far more "normal" than homosexuals.

A girl who has just started menstruating isn't nearly as fertile as she'll be a few years later, and her reproductive anatomy is still very delicate. Furthermore, there's some anthropological evidence that women who have their first child before the age of fourteen bear fewer offspring overall than those who become mothers in their late teens or early twenties. So although young females are high in "reproductive value" (in terms of the total number of their fertile years remaining), this evolutionary logic may not extend all the way down to gangly pubescent girls. Furthermore, the female's reproductive value is rather moot in this sense for a hebephile, given that, to use the modern example, he'd lose his attraction for the girl around the time she gets her braces off and grows out of her Justin Bieber fan T-shirts. Still, in the ancestral past, hebephilia may have represented an adaptive strategy under conditions where the risk of cuckoldry (unknowingly raising, and therefore investing one's resources in, another man's child) was especially high. After all, assuming she was at least reproductively able, the younger a girl was, the more likely she was to be a virgin, thus virtually guaranteeing the man's paternity if she became pregnant.* Depending on the most pressing adaptive problem for a man living in the ancestral past, different reproductive strategies varied in their effectiveness. And under such conditions of high paternity uncertainty (perhaps combined with high rates of STIs, as well, since youthfulness would also correspond with less prior exposure to sexually communicable diseases), impregnating, say, a dozen pubescent girls over the life span may have been a more adaptive strategy—at the heartless level of the man's gene replication motives only—than a monogamous man raising two children together with an adult woman. (Incidentally, if such an adaptive trait were heritable, it could help to explain recent genetic findings showing that identical twins raised *apart* are far more likely to share a hebephilic orientation than are fraternal twins raised *together*.)

*Such evolved motives have been portrayed unwittingly in many books and films, including the controversial movie *Pretty Baby*, in which a young Brooke Shields played the role of twelve-year-old Violet, a prostitute's daughter in New Orleans in 1917 whose coveted virginity goes up for auction to the highest bidder.

Yet even if hebephilia weren't evolutionarily adaptive under *any* conditions, in the history of our species, there have been far more babies born to ovulating thirteen- and fourteen-year-old girls than there have to men who have sex with men. So for social conservatives to draw from that insipid old argument that homosexuality is "biologically unnatural" and therefore "morally wrong" is essentially for them to say that sex with pubescent girls is "biologically natural" and therefore "morally right." And that's rather ironic, isn't it, given that so much of the fuel for today's pedophilia panic and antigay mentality lies at the fiery heart of the right-wing community.

It was due in no small part to these reproductive realities (well, that and the challenges of doing a Tanner scale classification for every child involved in a sex-abuse investigation) that the APA ultimately rejected a proposal to add hebephilia to the *DSM-5*, voting instead to keep only pedophilia as a mental illness. But it was an intense debate while it lasted. Ray Blanchard, who spearheaded the campaign to enter hebephilia into the diagnostic manual, used plethysmography to prove that many sex offenders incarcerated for crimes against children are in fact hebephiles (attracted mostly to pubescents) and not pedophiles (attracted mostly to prepubescents) and therefore presumably constitute a separate threat. Yet many of his fellow colleagues failed to see how simply being able to distinguish between pedophiles and hebephiles in prisons meant that the latter should also be diagnosed with a psychiatric disorder. One of Blanchard's most vocal adversaries was the forensic psychologist Karen Franklin, who accused him of confusing morals and science (much as the Victorian-era doctors had confused morals and science in the case of nymphomania, which similarly gave the courts license to confine individuals in mental hospitals against their will). Blanchard's bid to add hebephilia to the *DSM-5* was also met with resistance from the sexologist Richard Green, founder of the International Academy of Sex Research. Green titled one of his critiques of Blanchard's arguments with just a *hint* of transparency: "Sexual Preference for 14-Year-Olds as a Mental Disorder: You Can't Be Serious!!"

A sole attraction to pubescent girls is another thing altogether, perhaps, but in keeping with this amoral, mechanistic reasoning, and given that most of a woman's eggs are gone by the time she reaches the ripe old age of thirty, the *inability* to be attracted to young females who display visible signs of reproductive capacity (such as breasts and widening hips) would be decidedly abnormal (and I hope you're able to see clearly by now, by the way, why the issue of "normality" is so morally vacuous and why the question of harm must instead prevail before we can ever hope to make any real ethical progress in these debates). In the ancestral past, a man aroused more by women in their thirties or forties than he was by those in their teens or twenties would have been at a distinct reproductive disadvantage. Regardless of monumental changes since the Pleistocene days in our understanding of teenage emotions, our appreciation of older women, and the extension of the average life span, it's these youth-detecting ancestral men's brains that modern men still come standard equipped with. Age is just a number, yes, but as that number rises, the amount of a woman's eggs declines. That's not a sexist or ageist statement; it's simply a plain biological fact. (Note also that lust and love are wholly different; what inspires a man's lust will never change, but his love will adapt to whatever nature can throw at it, which certainly includes something as insignificant to an otherwise happy and successful marriage as his wife's menopause.)

None of this is to say that men who find women of more "suitable" ages attractive aren't exhibiting a biologically adaptive response. So long as the woman is still showing signs of being fruitful (whether by a genuine ripeness or false advertising by Botox), the capacity to become aroused by, say, those forty-something

women popularly known as MILFs (get thee to urbandictionary
.com if you don't know) is clearly biologically adaptive. It's just
not *as* adaptive as being attracted primarily to younger women.
As we saw in our discussion of parental investment theory, men
have more than enough (and then some) spermatozoa to spare,
so even if there's only the remotest chance of impregnating
an older female, the inability to be aroused by her could work
against a man's genetic interests. (It's basically the same principle
at work behind most men's capacity to be aroused by pubescent
girls, only it's applied to the other far end of the female reproductive age spectrum.)

This evolutionary logic is also why gerontophilia is so uncommon. An attraction to the elderly is apparently so rare, in
fact, that even Alfred Kinsey doesn't mention it in his otherwise exhaustive *Sexual Behavior in the Human Male.** (He
included plenty of other deviant sexual behaviors, such as pedophilia and bestiality, so it's a revealing omission.) Any man
whose erections are reserved for women over sixty, no matter
how lovely, active, and intelligent those ladies may be, isn't
much of a threat to other males in the genetic arms race (he'll
have a great sex life, though, with not only less competition but

*It bears asking whether there are meaningful cultural differences when it comes to
the sexual appeal of the elderly, particularly of elderly women. Writing of the Wogeo
tribe of New Guinea, the anthropologist Ian Hogbin came upon a woman advanced in
years who still actively seduced and had intercourse with many of the youths on the is-
land. "Desire doesn't disappear with the teeth," she told Hogbin. "And so long as a
woman can still dig she wants to do a little something now and then." Similarly,
Bronisław Malinowski, an anthropologist in the early twentieth century known for his
frank and unapologetic accounts of explicit sexual behavior among the natives of the
Trobriand Islands of Melanesia, traced an outbreak of venereal disease to a woman who
was, according to him, "so old, decrepit, and ugly" that nobody suspected her of being
the source of the epidemic. Such ethnographic data may imply a greater willingness to
have sex with the elderly in some cultures. But in support of the evolutionary account,
at least one survey has revealed that the most common cross-cultural reaction to the
elderly as prospective sexual partners is, by far, one of pronounced erotic distaste. See
Rhonda L. Winn and Niles Newton, "Sexuality in Aging: A Study of 106 Cultures,"
Archives of Sexual Behavior 11, no. 4 (1982): 292.

the appreciation, experience, and accumulated wisdom of his erotic targets).*

Wakefield's criterion of biological dysfunction aside, if it doesn't bother the gerontophile or his elderly partner (and by the way, many older women near and dear to me would wring my neck for calling them "elderly," but I mean no harm, I speak only of your absent ova), there's no reason for gerontophilia to be seen as a mental disorder. And indeed, you won't find it in the *DSM-5*. Its only dark manifestation would involve (as it does on some occasions) elder abuse.† But otherwise, it's kind of a win-win. Remember, the erotic age orientations are lifelong arousal patterns, and so a true gerontophile is a gerontophile at the age of twenty just as he is at the age of eighty. If such a man finds himself with a woman younger than he prefers, she's literally becoming more beautiful to him as she ages. From his subjective perspective, she's still merely larva slogging through the ravages of her youth throughout her thirties and forties, but by the time she's in her fifties, she's entered the dreamy chrysalis stage and will emerge as the most spectacular butterfly—entirely white and ever so fragile—at age eighty-five. That's the gerontophile's erotic ideal. You may see it as peculiar, but I suspect you'll find it less disturbing than the lamentations of a pedophile. "Little girls are hopeless

---

*John Money defined gerontophilia as "the condition in which a young adult is dependent on the actuality or fantasy of erotosexual activity with a much older partner in order to initiate and maintain arousal and facilitate or achieve orgasm." See John Money, "Paraphilias: Phyletic Origins of Erotosexual Dysfunction," *International Journal of Mental Health* (1981): 75–109. But neither Money nor Krafft-Ebing, who first described this condition, said how old *old* really was in the case of gerontophilia. In 2005, the psychiatrist Hadrian Ball suggested the clinical cutoff should be an erotic target aged sixty or more. Regardless of chronological age, it's the physical signs of advanced age that do it for the true gerontophile.

†In the U.K., for example, somewhere between 2 and 7 percent of all rapes involve victims over the age of sixty, and at least a subset of these cases are believed to involve a specific targeting of elderly victims. See Hadrian N. Ball, "Sexual Offending on Elderly Women: A Review," *Journal of Forensic Psychiatry and Psychology* 16, no. 1 (2005): 127–38.

causes because they grow into big girls in short order and then are unattractive," a team of sociologists quoted one such dreary figure. "That is why it is rare I know a girl for more than a few months or years. Once they grow into puberty, we slide apart and go our separate ways."

Pedophilia is a mental illness in the *DSM-5*, whereas geronto-philia is not. Both are quite clearly biologically dysfunctional, the former by having an erotic target (a prepubescent girl) who is much too young to conceive, the latter by having an erotic target (a postmenopausal woman) who is far too old to conceive. In both cases the individual's erotic age orientation is out of joint with the most basic mechanics of reproductive biology. Both, arguably, also involve erotic targets who are vulnerable physi-cally and often mentally as well. That one is included in the *DSM-5* as a mental illness and the other is not tells us that it's not adaptation-based logic about "normal" and "natural" alone that guides psychiatric opinions on which erotic age orientations to pathologize, and perhaps it's not even *only* about harm. Maybe there's something else at work, too. And given that society takes the psychiatric community's nod over what's "normal" and what's not and applies that to its treatment of minority sexual orienta-tions, it's worthwhile for us to try to get to the bottom of what that "something else" is, exactly. (If you doubt this, ask an older lesbian what it was like being gay when the APA still considered her mentally ill, then compare her response with what a young lesbian couple has to say about being gay in the United States today—but don't rush them, for crying out loud, wait until they're back from their honeymoon at least.)

If male gerontophiles are indeed about as common as female pedophiles, then perhaps they're simply not worth the effort of pathologizing. Establishing baseline rates in the population is a critical part of any psychiatric research given that one can only make claims about "abnormal" psychology by contrasting it with the known normal. But the challenge with getting accurate per-centages of the human population that could be characterized as pedophiles, hebephiles, ephebophiles, teleiophiles, and geronto-philes is obvious. If there were no shame today associated with being anything other than a teleiophile, scientists would have had these phallometric stats in the bank long ago. But outside of forensic samples, the only men likely to participate voluntarily in your study are confident teleiophiles who have nothing to con-ceal and therefore, save a touch of indignity, no reason *not* to participate. How many wary pedophiles, hebephiles, ephebo-philes, and gerontophiles will escape your attention, by contrast, is impossible to say, but rest assured, lines of eager men won't be snaking around the corners for your study.

There are also ethical challenges faced by the researchers themselves. Let's assume, for instance, that you somehow man-age to get a completely random sample, not sex offenders doing their mandatory plethysmograph testing, but men without any criminal history from the general population. Nonetheless, some of your subjects will inevitably score as pedophiles and show zero arousal to adults. If there are young children living in any of these men's homes, or perhaps there are even pediatricians or grade-school teachers among them, what would you do? All they've done, really, is respond physiologically on this artificial task, and erections aren't against the law. Don't forget also that you assured your subjects, as is standard practice for psychological re-search, that you'd keep their results confidential and anonymous. That's probably the only way you could get anything like this ran-dom sample in the first place. On the other hand, children may be at risk if you fail to disclose these men's erotic age orientation to the authorities. Given your knowledge of their desires, many people would blame you if any of them ever harmed a child. If you're an ethical scientist committed to protecting both the rights of your subjects and the safety of the public, it's a catch-22.*

---

*In a creative effort to address child sexual abuse, the German government launched a massive media campaign in 2004 to encourage self-identified pedophiles to reach out and get help from supportive professionals. Using a blitz of public service announcements—highway billboards, prime-time television commercials, full-page newspaper ads, spots

I suspect that someday in the not-too-distant future—your grandchildren may even live long enough to see it—the field of cognitive neuroscience will have advanced to the stage that a person's entire erotic profile (whatever appears behind the windows of the sex slot machine spin) will be accessible by a quick brain scan, perhaps without him or her even knowing it's been done (think airport security). Until those rather invasive days are here, we're unlikely to get any meaningful erotic age orientation statistics. Yet we certainly do know the way that basic evolutionary biology works, and while it doesn't tell us much about the number of true pedophiles or true gerontophiles there are in the world, except to say that both are probably fairly rare, we *can* safely infer from this theoretical knowledge that there are far more men out there who are attracted to legally underage minors than most of us would like to believe.

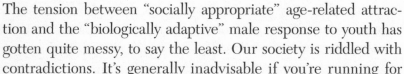

The tension between "socially appropriate" age-related attraction and the "biologically adaptive" male response to youth has gotten quite messy, to say the least. Our society is riddled with contradictions. It's generally inadvisable if you're running for mayor in your small town, for instance, but in Canada and several U.S. states one is perfectly free to debauch a consenting sixteen-year-old from dusk till dawn without fear of police inter-

---

before movie previews in cinemas—in the hopes of coaxing police-wary pedophiles out of the closet, the ad campaign read, "Do you like children more than you/they like?" (And notice the two meanings.) This was accompanied by images of coquettish children. "You are not guilty because of your sexual desire," it went on to say, "but you are responsible for your sexual behavior. There is help! Don't become an offender!" Over the next three years, 808 men responded to the ads. Of these, however, only 358 followed through with a face-to-face consultation with the psychologists in Berlin, so that's quite a lot of skittish pedophiles who slipped through the cracks after an ambivalent attempt to seek help. The ones who *did* follow through were committed to the project. Some had traveled all the way from Austria, Switzerland, and even England to volunteer. See Klaus M. Beier et al., "Encouraging Self-Identified Pedophiles and Hebephiles to Seek Professional Help: First Results of the Prevention Project Dunkelfeld (PPD)," *Child Abuse and Neglect: The International Journal* 33, no. 8 (2009): 545–49.

ruption, since that's the legal age where you live. Take a single naked photograph of your sex partner, though, and it's grounds for a federal sentence, since child pornography is defined at the national level as anyone under the age of eighteen.

We've become so uncomfortable about men's *possible* attraction to children that we've even started outlawing an aesthetic appreciation for the beauty of youth. In 2008, for example, the famed photographer Annie Leibovitz was commissioned by *Vanity Fair* to do a photo shoot of the Disney Channel's then fifteen-year-old starlet, Miley Cyrus. Some parents whose young children were fans of the teenager's Disney alter ego, Hannah Montana, were outraged by Leibovitz's shots: Cyrus was shown on newsstands across the world wearing smudged red lipstick, her long hair damp and tousled, with the length of her bare back to the viewer. She was clutching a satin sheet to cover her front, so it wasn't a topless image, but many nonetheless saw it as "suggestive." (My gay eyes just saw what her dad Billy Ray's "Achy Breaky Heart" had led to in 1992, hunched over and staring at me, but that's neither here nor there.) Online commentators called Leibovitz a "pedophile" and complained about the "sexualization of children" (incidentally, lascivious adults are hardly needed to turn a reproductively mature fifteen-year-old girl into a sexual entity; she's quite capable of doing that on her own), while Disney Channel representatives castigated the photographer and the *Vanity Fair* editors for "deliberately manipulat[ing] a 15-year-old in order to sell magazines." (This, you'll note, is a quote from the spokeswoman of a company that had built a billion-dollar franchise around that very same fifteen-year-old girl.) "I'm sorry that my portrait of Miley has been misinterpreted," Leibovitz responded to all the negative attention, clarifying that she'd discussed the shoot at length with a very agreeable Cyrus beforehand. "The photograph is a simple, classic portrait, shot with very little makeup, and I think it is very beautiful."

Another cultural icon to be skewered as a female "pedophile" is Germaine Greer. Her famous book, *The Female Eunuch*, has long been celebrated as a pivotal text in the feminist movement of the early 1970s. But it was her 2003 book, *The Beautiful Boy*, that caused the real ruckus. Essentially, *The Beautiful Boy* is a visual meandering along our undeniable history of eroticizing male pubescents in art (not many of Greer's critics were aware of the difference between pedophiles and hebephiles, needless to say). Featuring on its cover a photograph of the young actor Björn Andrésen taken during the 1971 film adaptation of *Death in Venice* (Andrésen had played the role of fourteen-year-old Tadzio, the beautiful Polish boy with whom the older male protagonist in Thomas Mann's classic novella becomes mortally infatuated), Greer's book showcases pretty boys frozen in creative amber, with artists like Caravaggio and Donatello immortalizing the je ne sais quoi of their youth before it slipped forever into the bloated or overly muscled bodies of piliferous men.* Greer describes her book as "full of pictures of ravishing pre-adult boys with hairless chests, wide-apart legs and slim waists."

Rather surprising to many, she'd actually written the book for women. In an editorial for London's *Daily Telegraph*, Greer clarified, "I know that the only people who are supposed to like looking at pictures of boys are a sub-group of gay men. Well, I'd like to reclaim for women the right to appreciate the short-lived beauty of boys, real boys, not simpering 30-year-olds with shaved chests." Despite her intentions, *The Beautiful Boy* indeed sold almost exclusively to that very subgroup of gay men—a glimpse, perhaps, into the relative scarcity not only of female

---

*Thomas Mann penned *Death in Venice* (something of the gay *Lolita*, or in Mann's words, "a case of pederasty in an aging artist") only after becoming entranced in real life by what Nabokov would call a "faunlet" while vacationing with his wife in Vienna. In the true story, the Polish boy whom Mann had become infatuated with wasn't fourteen but only eleven, and the child was later identified as the Baron Władysław Moes. See Gilbert Adair, *The Real Tadzio: Thomas Mann's "Death in Venice" and the Boy Who Inspired It* (New York: Carroll & Graf, 2003).

pedophiles but female hebephiles as well. When I asked a friend in her thirties if she could ever see herself getting turned on by a young pubescent male, I believe her response was something along the lines of "Have you ever *smelled* a fourteen-year-old boy's sneakers?" Clearly not a hebephilic podophile, that one.

In Greer's hometown of Sydney, New South Wales, a scandal erupted in 2008 when police raided the Roslyn Oxley9 Gallery in Paddington and seized more than a dozen figurative works by the acclaimed photographer Bill Henson. Henson has for decades been considered a national treasure by the creative class of Australia, his images gracing the walls of the Guggenheim Museum in New York, the Bibliothèque Nationale in Paris, and the Venice Biennale. He's made a career of focusing his camera lens on wistful adolescent subjects. In advance of the artist's Sydney exhibition, the gallery had sent out an electronic invitation to the New South Wales haut monde that included one of his new portraits of a naked thirteen-year-old girl. That image made its way to Hetty Johnston of the Bravehearts foundation, an Australian child advocacy group. A raid on the Roslyn Oxley9 was triggered when Johnston and other concerned members of the public lodged complaints with the police, in their view tipping off the authorities that the gallery was slated to feature a collection of child porn. After much ado, all charges against Henson and the gallery owners were dropped, but the incident resulted in a controversial revision to New South Wales child pornography laws that effectively removed "artistic purposes" from special exemption in the production of images of nude minors. The Henson affair sharply divided the Australian public, with enduring concerns over government censorship colliding headfirst with the pedophilia panic of the current age. Australia's prime minister at the time, Kevin Rudd, declared Henson's work "absolutely revolting" and of "no artistic merit." Art critics in turn rebuked Rudd as a philistine who was in no way qualified to offer such a critique. The gallery owners received credible threats that their building would be set ablaze. And the unflappable artist, for his part, was quoted a few years later as saying that the whole episode was "at best inconvenient."

To me, there's just something so very sad about this point we've come to: it's the point where a youthful nude in an art gallery no longer is seen as a thing of beauty but is instead only a tawdry broken mirror distorting our own terrifying desires. Children should be fiercely guarded and kept close to the breast of any civilized society. But in adopting a patently false but stubbornly clung-to mythology of human sexuality that makes demons out of natural drives, we've entered a stage of moral sickness, not of moral health. The good news is that it's just a stage, not a terminal illness. And as we've seen quite clearly throughout, when it comes to sex, we human beings are a work in progress.

SEVEN

LIFE LESSONS FOR
THE LEWD AND LASCIVIOUS

Indeed, of all the kinds of decay in this world, decadent
purity is the most malignant.
—Yukio Mishima, *Confessions of a Mask* (1949)

Modern cruise ships are floating metropolises with enough activities to keep a twenty-first-century kid with a Red Bull addiction and attention deficit disorder busy forever. But in the first
half of the seventeenth century, when a motley crew of British
expatriates outgrew their welcome with the Church of England
and decided to colonize America, being an émigré on one of
those more austere, rat-infested transport vessels—and stuck
with a bunch of Puritans for fellow travelers, no less—was dread-
fully dull. Imagine what it must have been like as a teenage boy
aboard the *Talbot*, for instance, which set sail from the Isle of
Wight in March 1629 bound for the newly founded Salem in the
Massachusetts Bay Colony. (This was long before the witches arrived, so aside from rumors of folks having sex with pigs, which
was nothing new, really, it was still the promised land over there.)
As an adolescent male, you're basically an ambulant sperm factory
with an incompetent foreman, but somehow or another you'll
have to get through months on this leaky, oversize bathtub with

only psalms and daily gum checks for scurvy to pass the time
and keep you from sinning. And that's easier said than done.

We know from the records that there were at least five such
boys in the belly of the *Talbot* on that journey to the New World,
because they're the main players in a landmark event in early
American history. It's not a tale you'll find in any public-school
textbook, but it belongs in there every bit as much as that of
Pocahontas and John Smith (and there's probably more truth to
it, too). When the *Talbot* finally arrived along with the rest of the
"Higginson Fleet" on the banks of Salem the morning of June
19, the most urgent order of business for the fleet master, the
Reverend Francis Higginson, was to deal with these "five beastly
Sodomiticall boyes [who'd] confessed their wickedness not to be
named" while aboard the ship. They weren't the first seafarers to
while away the long hours this way, nor, certainly, would they be
the last British schoolboys to do so, but these lustful teenagers
were the first recorded sodomites ever to set foot on what would
eventually be U.S. soil. (That's only the written history, of course.
There's no doubt some randy Native Americans beat them to the
homosexual punch eons before.) Their feet wouldn't be planted
on this side of the Atlantic for long, however. The details of how
it all came to light aren't clear, but we do know that once the
Massachusetts authorities learned of the boys' gay orgies aboard
the *Talbot*, they were so horrified that they shoved the teens
right back onto the ship and returned them to England, along
with a message to King Charles that since the unmentionable
crimes occurred on the high seas, *technically* these "beastly" lads
should be dealt with at their point of origin.

I was born some 345 years, 10 months, and 17 days after the
Talbot first came ashore at Salem, just a few hundred miles south
in Nowhere, New Jersey. It's not entirely clear if I was literally
born gay—it's not as if any of us are rushed into neonatal plethysmography straight out of our moms' vaginas—but whether I
was queer from conception or arrived with a brain genetically

predisposed to getting stamped with an irrevocable orientation to penises during my first few years in America, our country's initial planks of religious scaffolding had by then grown into a fortress of self-righteousness. Many Europeans like to point out that Americans have a "complex" about sex, but remember, if you go back far enough, we're European. It's just that our earliest settlers were among Europe's biggest prudes. Fortunately, the anti-gay sermons had eased up *slightly* by 1975. Yet in many ways, I slipped out of fetal solitude that year and into a society playing the same old broken record of thou-shalt-nots that Higginson and company had imported from an overly ecclesiastical Britain in 1629. After all, they might have sent that first wave of sodomites back to England, but the Puritans already setting up shop in the colonies carried all the necessary ingredients for deviant sexuality. And that was enough to keep the pitchforks sharp and the stakes burning for at least four centuries to come.

America has long had its issues with sex, but few modern countries haven't. One doesn't typically think of "gay Paris" in the 1920s, for example, as being a hotbed of oppression for homosexuals. And it wasn't, if by that you mean the sort of blind hatred toward gays and lesbians seen in other parts of the world (Uganda comes to mind today, with its backward "Kill the Gays" bill on the table recently). But even in cosmopolitan Paris, homosexuals were specimens before they were human beings. If you were a gay man looking for a state-of-the-art solution to your "condition" of sexual inversion, then Paris in the early twentieth century was the place to be. There you'd find doctors singing the praises of a promising new experimental treatment in which your testicles would be replaced with those from the cadaver of a hypersexual straight man. It didn't work, by the way, and the straight French convicts implanted with their recycled testicles didn't turn flamboyantly gay either. Luckily, however, they all got

human gonads. Over in Spain around this same time, a few ren-
egade doctors were busy grafting monkey testicles onto their gay
male patients.* Even more bizarre, they were grafting just a solitary monkey testicle somewhere on the man's body. Rather curi-
ously, the records don't say where on his body it went, exactly.
"What in the world is *that*?" I can imagine one of these men
hearing while undressing before a new boyfriend years later.
"This lump on my back? Oh, it's nothing really, just one of those
monkey balls from when I was young and stupid."

We now know that these physicians who viewed homosexuality as an endocrine problem were barking up the wrong tree. But
at least their approach shows a shift toward scientific thinking
about gays and lesbians. That thinking was pretty shaky and obviously inhumane, but nonetheless it was probably an improve-
ment on the long-standing superstition and fear hovering over
the subject of sodomites. Not everyone, of course, embraced rationalism; even today, many of us are stuck in the puritanical
mud of 1629. But the opportunity to pull oneself out of the residue of such fire-and-brimstone reasoning is available to anyone
wishing to reach out and grab hold of the best science of the day.
Science won't tell you what's moral and what's not. But by standing up on your own two feet on the terra firma of reality instead
of remaining up to your eyeballs in the swamp of dogma, you'll
get a much better lay of the land for navigating the moral landscape. And if you head one day in the wrong emotional direction,
you'll know that more than likely, it's just the gunk from an overly

*Leading the charge for this monkey-testicle cure for male homosexuality was a medi-
cal professor from the University of Madrid named Gregorio Marañón. In his *Evolution of Sex and Intersexual Conditions*, published in 1930, Marañón writes: "Several
[physicians] have endeavored to combat homosexuality by replacing the testicles of the
invert by those of a healthy man; or by grafting upon him the testicle of a monkey . . .
with results that are favorable, though still subject to criticism." He confesses, however,
that he's had a hard go of finding these "favorable" results in his own experimentation:
"[In] one of my own cases of homosexuality, the grafting of the testicle of a monkey
augmented the libido, but in a homosexual direction" (168–69).

religious past clogging the inner workings of your moral compass, steering you away from "against what is right."

A purely scientific approach to sex, however, especially one that trades exclusively in the language of the "natural" and the "normal"—and one in which the word "harm" either never appears or is never properly defined—can send us scampering off to a this-worldly Hell as easily as a religious approach. We've seen, for instance, some of the unintended consequences of treating sexual deviance from a purely medical perspective, especially how the practice of pathologizing minorities has in many ways done more damage than good (both to the minorities and to the rest of us). Just look at all those men who went to their graves in Paris with someone else's family jewels sewn into their scrotums. As we learned in the first chapter, once researchers began to understand erotic orientations to be lifelong patterns of attraction, human beings became distinguishable from one another not just on the basis of, say, skin color, nationality, and social status but also on the basis of their main turn-ons. For most people, this new concept of "orientation" in the latter half of the nineteenth century was an inconsequential development. But it was a change that would strike fear in the hearts of many others forevermore; after all, with experts now separating the "normal" people from the sex deviants, getting exposed as one of *those* people came with all sorts of problems. It wasn't just a medical diagnosis; it was a social sentence. As a consequence, modern societies became giant breeding grounds for a whole new oeuvre of shame- and anxiety-related psychiatric disorders. From that point on, being a human being with whatever erotic profile (or profiles) your society happened to hate the most would be like living permanently in Middle America as a Communist during the McCarthy era. Only in this situation, you couldn't just tear up your membership card to your socially inappropriate club if the stress got to be too much; your membership card was your brain.

If you were one of those homosexual specimens trying to

avoid notice back in Paris, on your trail would be investigators such as Professor Charles Samson Féré. This no-nonsense hetero-sexual was a physician who'd been inspired by Havelock Ellis's *Sexual Inversion*. While his colleagues were busy swapping testicles in labs along the Seine, Féré set out to develop a fail-safe method for detecting gays and lesbians who were concealing their homoerotic tastes. Kurt Freund wasn't even a twinkle in his mother's eye and that wacky erection-detection machine of his just some state-of-the-art piece of equipment for the faraway Jetsonian future. So all that Féré could do to find out who was gay and who wasn't, really, was try to work out the physical, behavioral, and psychological secrets of queers. In his *Scientific and Esoteric Studies in Sexual Degeneration in Mankind and in Animals*, the author shares the wisdom of this new gaydar, circa 1899. "Posture, demeanor, methods of walking," wrote Féré, "all partake of inversion." In a subgroup of gay men, he believed that certain bodily traits betrayed the patient's homosexuality. For instance, "there have been noticed the development of fat in the mammary parts, the large size of the buttocks, and scarcity of [body] hair" (okay, fine, guilty as charged on two of those, but so is my straight brother). And just in case you were wondering, gay men's penises look the same as straight men's penises. "A doctor had [dealt with] more than 600 homosexuals," Féré assures us, "without meeting with a single case of malformation of the genital organs among them."* Burning the midnight oil on many a lonely online night, I have reached my own sample size of more than 600 gay men's penises—hard to say precisely how many; I stopped counting once I hit a million back in '02—and I can confirm Féré got this point right. (Well, more or less. I'd be lying if I said I haven't seen some real doozies.) Don't fret, though, ladies, because while you may not be able to tell they like other

---

*Féré believed that although the genital organs of gay men looked normal in appearance, they *acted* differently: "In most cases there is irritable weakness. Orgasm often occurs with them as the result of a mere touch, of the sight or odor of the one they love" (146).

guys by inspecting their packages, Féré lets his readers in on a little secret about homosexual males, at least those of the more obviously inverted countenance: they find it hard to blow. Now I know what you're thinking ("Obviously, he hasn't met my friend Mark," or some such), but it's the "inability to learn how to whistle," Féré clarified, "that is the mark of the effeminate man."* Four words, Dr. Féré: *Clay Aiken, whistling fiend.*

Lesbians didn't escape Féré's investigations, either. Indeed, he was just as motivated to pull his gaydar dragnet over these more slippery sapphic properties. The doctor was convinced—in a rather Freudian sort of way—that lesbianism was caused by a girl's obsession with her mother's breasts in early childhood, and she became jealously enraged upon seeing her father paying special attention to, or even daring to touch, what she considered to be hers alone. This animosity toward her boob-poaching father, claimed Féré, was the impetus for a lesbian's lifelong disgust for the opposite sex. Still, he figured there must be some biological predisposition for a girl reacting this way. "The sight which [shocks] her is such an everyday affair that one might almost say that there is no child who has not seen it . . . the acquisition of an instinctive perversion on account of it could only take place as a consequence of a special aptitude for such acquisition." Nonetheless, for the rest of her days, the girl would now view men as mammary thieves, with her most intense passions directed at bosomy females.† The best gaydar tool for detecting closeted lesbians, therefore, is to watch carefully where their eyes go in

---

*Gay men "practice gymnastics," the doctor added to his list. "They do not give way to tears. As a rule, they like dancing, but with persons of their own sex . . . the effeminized man especially likes coachmen, butchers, circus-riders, etc., or persons who have large sexual organs" (154).

†Féré describes one of his own lesbian patients with this background, coming of age in France: "She was attracted by girls and felt an urge to caress them . . . she noticed the rubbing of her breast against theirs caused specially pleasing sensations. When she was 16, she felt for the first time her genitals sharing in the excitement, becoming wet. From that time she began to have voluptuous dreams in which girls always played the most important part" (222). Even after she was introduced by another lesbian to what Féré calls "the mysteries of vulvar rubbings," it was always breasts that did it for her.

the presence of a shapely female. If she stares at the woman's voluptuous chest, then you can be certain her jig is up.

Most of his thinking about gay men and lesbians, as you can see, was just plain silly, but Féré also spread some pretty pernicious stereotypes about the moral character of homosexuals. Or rather, their lack of any such character, in his view. "It should be remembered," he admonished his readers, "that inverts have a tendency to lying, are vain, garrulous, and indiscreet. Some pay no attention to the dress or even the cleanliness of those whom they are on the lookout for. The most squalid creatures do not repel them." Oh, come on, with that last point. We're not *all* Lady Gaga fans, for heaven's sake. In any event, Féré's words may read to us today like a passage from the Westboro Baptist Church's monthly newsletter, but bear in mind, this is from one of the most respected physicians and scholars of his day. And in the history of our species, that wasn't so long ago. Even the gay-friendly Havelock Ellis hailed Féré's book (which was published only two years after his own *Sexual Inversion* appeared in English-language bookstores) as "the greatest work on the sexual instinct written in French." (This leads me to believe that Ellis's French wasn't very good.)

Féré also waved his finger in disapproval at perverts of other species, regaling his readers with shameless tales of masochistic stallions, a donkey with a passion for zebras, inverted hens, masturbating weasels, and even pederastic insects. There's no question that other animals sometimes engage in sex practices that are unusual within the context of the standard mating behaviors of their own species. Humans aren't the only sex deviants in the animal kingdom. But we *are* the only ones to stigmatize each other as disgusting perverts. To understand why we're so unique as a species in this unfortunate way, why so many intelligent people have fought with each other for so long over the very types of issues we've seen during the course of this book, we'll need to rewind the clock far beyond any recorded history, all the way back, in fact, to the time when we *Homo sapiens* became the insufferably judgmental hominids that we are. In doing so, we may be able to finally comprehend this existential sexual mess that we've come to find ourselves in—and why Féré's penchant for stereotyping sexual minorities is something we still haven't shaken.

❧

As a species of primate, we are principally set apart from other animals by our highly developed social cognition. Like a bat's radar system, which enables it to navigate through a dark cave or to find a crunchy insect to snack on, or like an elephant's trunk that allows it to do everything from snorkeling under water to nudging the "little" ones along on a family outing, our species's most distinctive adaptation is our ability to empathize—to think about what's going on in another mind. It's intuitive, it's innate, and basically we can't help it: we're constantly trying to get into the heads of others.*

In technical terms, this adaptation is a social cognitive mechanism known as "theory of mind."† The researchers who coined this term in the late 1970s, the psychologists David Premack and Guy Woodruff, did so to refer to the fact that minds are, by definition, purely theoretical. A neurosurgeon in the middle of an

---

*Scientific opinions are mixed as to whether we're *entirely* unique in having a theory of mind or whether we just can't detect it as readily in nonverbal animals (a few other social species, such as great apes, dogs, dolphins, and crows, may have some ability to reason about other minds also). But there is a general consensus among researchers in this area that humans are the planet's "natural psychologists." This doesn't make us "smarter" or "better" than other species—you wouldn't say that a bat's radar abilities make it smarter or better than an elephant with a trunk, after all—just different. See Mark Nielsen et al., "Social Learning in Humans and Nonhuman Animals: Theoretical and Empirical Dissections," *Journal of Comparative Psychology* 126, no. 2 (2012): 109–13.

†The word "theory" here refers not to a formal academic theory but to the default human way of perceiving other minds in the world. It's not fully functional at birth, but it develops rapidly over the first few years of life, with studies consistently revealing that a theory of mind is up and running by a child's fourth birthday.

operation isn't actually looking at a *mind* on the table; he's looking at an organ that produces mental states in its owner. If we
were to pore over some batch of fMRI results or recordings from
an EEG, the images we'd have before us come from "brain-
imaging" devices, not "mind-imaging" devices. That's to say, regardless of what some eccentric characters out there may try to
tell you—or rather to sell you—we can't literally see, hear, touch,
taste, smell, or otherwise directly perceive a thought.* Instead,
we can only theorize about other people's mental states (hence,
"theory of mind"). So whether you're a cognitive neuroscientist
in a lab deciphering blood-flow changes in an epileptic's brain or
just a guy on a busy sidewalk trying to make sense of a street
vendor's perplexing behavior, you're still only theorizing about a
mind. Other minds *do* exist; it's just that, as with gravity, we can
only infer that they're present on the basis of what we directly
perceive through our own sensory organs.†

Now, just because we come with this theory of mind system
factory installed doesn't mean that the specific theories it gen-
erates concerning what's going on in someone else's head are
always correct. Since we can never know everything about a
person's private mental life, more often than not we only get it
partially right. And a lot of times, we get it flat-out wrong. Imagine, for instance, that you're riding on a crowded subway. You're
somewhat oblivious to your surroundings given that you're busy
texting and wearing headphones (perhaps you're even whistling
away to your favorite song, assuming you're not an effeminate
gay male). Seemingly out of the blue, the guy across the aisle—a
real bedraggled sort, probably headed to the homeless shelter,
you told yourself when he got on a few stations ago—suddenly
lunges aggressively at that well-dressed businessman with the

---

*Which is why sexually deviant thoughts are, in and of themselves, inherently harmless.
†Unless you want to adopt the extremist philosophy of solipsism, which posits that since
we can't directly perceive them, there's no reason to assume that other minds exist at all.

charming smile and salt-and-pepper hair whom you also noticed before. With all the chaos (people scrambling to get away, disheveled newspapers sailing from laps, shrieks and cries) you're not going to pull out your pen and legal notepad from your bag and calmly sketch out a theory about what in the hell is going on; instead, your evolved theory of mind system has automatically, effortlessly kicked into gear. So, tell me, why *does* the angry unshaven man with the flies circling around him have his hands around this other gentleman's neck?

If you're like most of us, your knee-jerk assessment is that the assailant is clearly mentally unstable. But wait, what's this? A distraught woman clutching her teenage daughter is shouting obscenities at the handsome assaulted executive (actually, he's a little less handsome now with that broken nose of his). She's accusing him of—and you can hardly make it out with all the commotion—touching the girl while her back was turned. Ah, so *that's* it. The businessman wasn't so innocent at all, it seems; he's a practicing frotteurist. With that informative update, watch as your theory of mind transforms the "filthy, disturbed assailant" into a "down-on-his-luck hero" before your very eyes. He may not be a knight in shining armor so much as one in rags of stale urine, but he's one of the good guys now to you regardless.

The evolution of theory of mind was a huge boon for our ancestors. The more information they had about what was happening on the flip side of another's skull (his intentions, his desires, his emotions, his knowledge, his beliefs, and so on), the better our ancestors' "guesses" about why the other person behaved as he had and, even more important, what he was going to do next. (As my graduate adviser liked to say years ago, "The best predictor of future behavior is past behavior.") Being able to explain and anticipate actions like this was a game changer for a highly social species like ours. And when it comes to how this uniquely human adaptation affected our sex lives, the consequences of being able to think about what others think were massive indeed.

•

Consider what your average sexual encounter would look like in the *absence* of a theory of mind. Here's the completely "mind-blind" straight male's perspective, for instance, on entering an everyday domestic bedroom scene: A large object of a uniform pallor, pointy caps at the ends of two compact swellings, is twisting about on the sheets. A pair of thin stalks that balanced the object when it was upright have now drawn apart to reveal the soft pink interior of a woolly black diamond. Something red has poked through a different opening up top; it appears to be a papillate organism moving back and forth through a ragged white gate. Meanwhile, a set of restless blue marbles has settled in place above, and if one looks closely, inside these marbles are black dots exhaling like small resting ravens.

This is nothing at all like what husbands see when they stumble happily upon their ready-to-go wives in bed (at least, I hope it's not). But without a theory of mind that enables them to perceive the "object" as a conscious human being like them, it's indeed, horrifyingly enough, something like how a heterosexual man would be processing this sensual scene. Although the most common assumption is that seeing people in the buff or wearing revealing clothing serves to "objectify" them in our minds, the real effect on our social perceptions of seeing a bounty of flesh is quite to the contrary. Or, as it usually is with science, it's a little more complicated than that. In 2011, the psychologist Kurt Gray investigated what it is, exactly, that we see in other people when they're not wearing their clothes. In one of several such experiments, Gray and his colleagues had 527 participants, men and women with an average age of thirty-one, take a good long look at photographs of attractive models (also of both sexes). Participants were asked to rate each model on his or her general psychological competencies: "Compared to the average person, how much is this person capable of *feeling joy*? of *planning*? of *self-*

control? of *feeling pain*?" and so on. The models were either completely nude or fully clothed, which was the only difference between them—pose, lighting, and facial expressions were otherwise identical.* Although the nude models were judged to be less capable of what you might call "intelligent thought" (basically, anything involving executive functioning) than the clothed models, they were judged to have a far greater capacity for experiencing physiological or emotional types of mental states (such as pain, hunger, pleasure, fear, desire, rage, and joy). "The idea that [nudity] can lead to both decreased and increased mind stands in contrast to the term 'objectification,'" Gray explains. "Focusing on the body does not lead to de-mentalization but to a *redistribution* of mind." The effect was even more dramatic when the naked models were shown in sexual poses similar to that of the "objectified" wife in our scintillating boudoir episode (you remember her, the one with the "woolly black diamond" for a vagina and the "papillate organism" for a tongue?).

Such a social cognitive distortion, in which the image of another's abundant flesh blurs his or her intellect in our mind while sharpening our focus on his or her naked *feelings*, translates to our actual sexual behaviors toward this other person as well. We're not usually preoccupied with our partners' math skills or flair for languages as we're screwing (or being screwed by) them. But we're highly attuned to what they're experiencing at a more sensory and emotional level. Far from objectifying our sex partners as slabs of meat, we're very much aware of the pleasure or pain that we're causing them. Even a sexual sadist doesn't view other people as objects. In fact, it's the other way around altogether. The sadist is able to derive pleasure only through the lens of his theory of mind,

---

*Each participant judged each model only once, but for every model there was a nude image and a clothed image that was identical aside from the amount of flesh shown. (In other words, all of the participants got a mixed batch of nudes and nonnudes, but no participant ever saw the same model both naked and clothed. It was one or the other.)

in this case by perceiving a mind capable of experiencing the pain he inflicts. His sadistic arousal is inflamed by the very "redistribution" of mind described by Gray; with his erotic target now stripped of any significant cognitive functions, what he sees before him is a tingling, hypersensitive, wide-eyed figure whose entire axis shifts with every cruel touch he makes. For those of us with slightly less frightening sex lives, the underlying mechanism is the same; it's just that our arousal is titillated by the perceived pleasure we're inducing in others and not by their pain.

In fact, for a loving couple to be able to synchronize their sexual movements so expertly that their orgasms can be delayed for an hour or more, with one member releasing his or her own restrained passion so that it erupts in tandem with that of the other, requires very advanced social cognition. This isn't going to be our experience with *every* erotic encounter we have (really, who has the time), but with some basic practice under our belts, or at least some penis-numbing prophylactics, it's within erotic reach of most of us. Even our three-minute coital average is impressive when compared with the mad fifteen-second gnashing of genitals in our closest primate relatives. Whether we've mastered the *Kama Sutra* or prefer a quick splash in the gutters on Lovers' Lane, human sex is almost always an elaborate dancing of loins, an intersubjective ballet of lust. Unfortunately, we can still never *really* "become one" through even perfect climactic timing. Don't forget, after all, our sex partner's mind is there in theory only, an especially sad solipsistic fact that led the poet William Butler Yeats to write so ruefully of the "perpetual virginity of the soul."

Theory of mind might not allow us to literally penetrate another person's subjective existence (maybe it's for the better anyway—it sounds like something that would take months to clean from the walls), but it does animate that person in a way that shows us such a presence dwells beneath the skin. Once this social cognitive system evolved, it became clear that others were psychological entities like us, with sexual desires of their own. More important, we could see that those desires weren't always so neatly aligned with ours. The most positive sexual consequence of this psychological innovation is that it enabled us to conceptualize the mental construct of "consent." It's a bit nonsensical, for instance, to use terms such as "rape" or "sexual coercion" to refer to behaviors in other species in which a male copulates with a female struggling to free herself, and there are, indeed, many species in which such a pattern is common. The male simply doesn't have the evolved social cognitive equipment allowing it to think of the psychological harm that it's causing to the sex "object." Imagine a monkey in the act of thrusting saying to itself, "You know, actually, wait, I wonder if she's comfortable with me sticking my penis inside her like this?" By contrast, men who violate females (or other males) this way are indeed rapists. Assuming the other person is giving clear signals (the word "no" is usually a pretty good one), a rapist's theory of mind allows him to detect the mental state of unwillingness, and yet he continues with the act anyway. This unique ability to ascertain the other individual's psychological consent was the cognitive key needed for unlocking any coherent form of sexual morality in our species. While human societies are far more different in their attitudes toward sex than they are similar, and the form and degree of punishment for sexual transgressions vary enormously, no known culture on this earth has ever smiled upon rape among its own citizens.

Being able to reason about another person's thoughts also brought with it a strange, and sometimes disconcerting, mental effect in our species: the feeling of sexual shame. By using our theory of mind to take the mental perspective of someone else, we were able to see ourselves as he or she saw us. That could be a rather unflattering sight when it comes to sex. Just as their attention could turn to someone else's erotic motives, our ancestors became cognizant of the fact that another person could speculate on their desires. This led to unspoken rituals of sexual deception. It can get quite Machiavellian, but in one of its simplest forms, if you've ever had a crush on someone whom you didn't want to know about it and so you deliberately hid those fire-in-your-pants feelings, you've engaged in such theory-of-mind-driven deception.

Related to this is another unpleasant reality: We may desperately *want* to be seen as sexually desirable to someone else, since that's how we feel about the particular person, but unfortunately we're just not his or her type. (Trust me, few know unrequited love better than a gay man.) Yet as the trillion-dollar cosmetics industry attests, that definitely doesn't stop us from trying. On the other hand, being intensely desired by someone toward whom we feel no attraction at all can also be disconcerting. It's not merely finding out that someone you don't really fancy has a harmless crush on you. That may be. But there's also a distinctively unpleasant phenomenology (or the *what-it-feels-like* sense) that comes from knowing that your body is inducing an intense degree of sexual arousal in someone you'd actually prefer it didn't. This is precisely the state of mind that many feminist writers are referring to when they use the word "objectification," or when they define porn—aptly so—as the "articulation of the male gaze." Here's how the author Angela Carter describes this peculiar feeling of being someone else's erotic target in her short story "The Bloody Chamber" (which is the one for you if you're a man who'd like to know what it feels like to be a woman but you're not so committed as to invest in a whole new wardrobe):

> I saw him watching me in the gilded mirrors with the assessing eye of a connoisseur inspecting horseflesh, or even of a housewife in the market, inspecting cuts on the slab. [The effect] was strangely magnified by the monocle lodged in his left eye. When I saw him look at me with lust, I dropped my eyes but, in glancing away from him, I caught

sight of myself in the mirror. And I saw myself, suddenly, as he saw me . . . the way the muscles in my neck stuck out like thin wire.

"There are some eyes," writes Carter, "that can eat you." Of course, a person's sexual subjectivity, as we've seen throughout, complicates these matters even more. After all, an exhibitionist revels in this very notion of being consumed by others' eyes. (And in fact Carter's own empowered female characters often find their loins stirring unexpectedly at the thought of their trembling "horseflesh" being held and rotated in a man's carnivorous mind.)

Beyond our own private liaisons, our evolved theory of mind system also enables us to morally evaluate (as we've been doing all along in this book) those whose sexual natures differ so drastically from our own. And when this system isn't held in check by scientific facts, our impulsive judgments of these erotic outliers can be heinously harsh. Much of the trouble in this area stems from the fundamentally egocentric nature of our social cognition. I can no more reliably take the perspective of a middle-aged straight man aroused by the sight of a woman's genitalia, for example, than I can that of a male hamadryas baboon getting worked up over the amorphous, rainbow-colored swelling on the calloused rear of his female lover. (I mean that, for better or worse. If it's not perfectly apparent to you already, I'm as gay as they come, a "Kinsey 6," you might say.) Yes, understanding reproductive biology enables me to think logically and mechanically about such heterosexual cues. But metaphorically speaking, having to slip into either of these male primates' skins isn't the most pleasant form of virtual reality for my gay human brain. And as we saw earlier, when we're blue around the gills, our moral reasoning abilities aren't exactly at their sharpest.

Let's flip this example around and see what happens when a completely heterosexual man (a "Kinsey 0" on the zero-to-six scale) is told to imagine having sex with another man. In a 1979 study by the psychologists Donald Mosher and Kevin O'Grady, straight college guys were shown clips from gay male porn and instructed to identify with one of the actors in the film: "[Experience] the emotions that you would have if you were, indeed, engaging in the sexual behavior." The result, as you'd guess, was disgust, anger, shame, contempt, and greater agreement to such eloquent survey items as "I've never been able to understand why anyone would fuck a man in the ass when you could have better sex with a woman"; "You can't walk into a men's john these days without some guy looking at your cock or showing his hard-on"; "I'd rather be dead than queer"; and "You can tell a pansy by the flowers and butterflies that he wears."*

Fortunately for both fashion and gay rights, the 1970s were laid to rest under an orange-and-brown linoleum floor somewhere decades ago. But although their exact contents may be different, the brains of college students today work pretty much the same as the brains of those in 1979, just as *their* brains worked the same way as those of the eighteen- and nineteen-year-olds who lived millennia before them. Natural selection is an incredibly sluggish business and doesn't move at anywhere near the lightning pace at which human knowledge accumulates. This is a vital point in the context of this discussion, because until our species evolves a totally new kind of brain, any moral progress made toward the subject of sexual diversity hinges

---

*Male participants who saw scenes of gay anal sex not only experienced more negative emotions than did those who watched a video of a solitary man masturbating but also reported more "genital sensations." The authors interpreted this to mean that straight men's anger and disgust toward gay men are reactions to their own suppressed desires. More recent work by the psychologist Henry Adams traded in self-reports for data from a penile plethysmograph. Adams found that, basically, the more hostile a man is toward gay men, the stronger his erectile response is to gay male porn. See Henry E. Adams, Lester W. Wright, and Bethany A. Lohr, "Is Homophobia Associated with Homosexual Arousal?," *Journal of Abnormal Psychology* 105, no. 3 (1996): 440–45. An alternative interpretation does exist (essentially, that anger toward gay men induces a more general physiological arousal causing an erection). I'll let you judge for yourself on this one.

solely on the use of our acquired knowledge to defuse our cru-
eler, instinctive biases.

In the modern world, which is a land where entire lives are
tidily reduced to a letter on a string of ciphers ("LGBTQ" and
whatever else gets thrown onto such messy sandwiches of community acronyms), it has become more imperative than ever for
us to resolve this terrible tug-of-war between our innate judgments and our critical-thinking skills. With human beings carved
up into so many sexual "types" (and subtypes), negative *stereotypes* will spread over them like some invidious algae. If these
continue to grow for too long without anyone putting a stop to
them, it will become all but impossible for us to make out the actual human being—the individual person—beneath. In fact, that's
exactly how it all evolved to work. Negative stereotypes develop
immunity to moral logic because they have an undeniable adaptive currency. Our brains systematically collect and aggregate all
the negative information they can about the most salient categories of people in our social environments. Since we can never
meet every member of every category, the unpleasant tidbits gathered by our brains come from a very limited sample only. Yet that
doesn't keep these prejudiced organs of ours from automatically
and unconsciously—and often against our own better thinking—
ascribing these undesirable traits to everyone in that demographic.

Take our heroic homeless man back on the subway, for example. Which of the following was in fact the safer assumption? (And
before you answer this, remind yourself how you were traveling
at a high rate of speed beneath the surface of the earth in a confined vessel at the time, and so you couldn't exactly run away to
safety as the incident flared up.) Was it that the homeless man
had psychiatric problems making him dangerously unpredictable, or that the silver fox in the thirty-five-hundred-dollar tai-
lored suit must have done something nasty to provoke the attack?
It's wonderful, really, that your negative stereotype of homeless
people as being mentally unstable was so fantastically wrong in

this case, but your negative stereotype was still "right" in the amoral sense of leading you to err on the side of caution for your own selfish genetic interests. (You may be all winks and smiles with the chivalric transient now, but had that mother never screamed about her daughter, you'd still be diligently avoiding any eye contact with him.) As I mentioned briefly in the first chapter, this better-safe-than-sorry function of stereotyping helped our ancestors to make the best split-second decision possible with only limited social information to go on. But it also turned us into ready-made bigots. With our biased attributions made possible by our theory of mind, we simply expect the very worst in strangers.

By stereotyping individuals due to their sexuality—the "lesbian," the "transvestite," the "pedophile," the "fetishist," the "exhibitionist," the "masochist," and so on—we've lost the trees for the forest. The reason our knowledge of a person's hidden sexual desires overshadows everything else we know about him or her becomes clear in the context of evolutionary theory. At their core, of course, adaptive behaviors are those that aid an individual's reproduction, and so it's hard to imagine having any more useful, or strategic, information about a person than the nature of his or her sexual desires. Aside from the fact that it tells you I'm not an adventurous person when it comes to Asian cuisine, for instance, I doubt it would interest you to know that I had a humdrum plate of chicken pad thai last night. But if I told you that after dinner I *finally* lost my heterosexual virginity to a stunning, and unusually patient, given the circumstances, Thai waitress in the restaurant lavatory, I suspect your ears would perk up a bit more. (And on that example, remember what your mother told you: "If it sounds too good to be true . . ." Unless I get a brain transplant, and therefore I cease to be me, I'm afraid this particular penis will never see the inside of a vagina.)

People have no control over their sexual orientation, but likewise we have no control over the fact that our brains evolved to pay special attention to and systematically accrue information about other people's sexuality. Again, from an evolutionary perspective, those details are high-dollar knowledge. So let me come clean. The truth is that the "Thai waitress" from last night was in fact a married man sitting with his wife at the next table over. After he made eyes at me for the better part of an hour, we met in the restroom and exchanged a brief but passionate moment in one of the toilet stalls. Oh, I'm still pulling your leg. But you see how sex stirs the social brain. Flitting quickly through your head a moment ago were probably incoherent thoughts such as "Hold on, did the man's wife know what he was doing?" and "Where was Juan during all this?" The critical takeaway here is that while we can't undo natural selection and reengineer human social cognition so that we've no interest in other people's sexual desires and behaviors, we have considerably more control over what we do with that information once it's been revealed to us and how we treat a vulnerably "exposed" person as the result of our knowing. Like fighting alcoholism, the first step in overcoming our sexual bigotry is recognizing that we're sexual bigots.

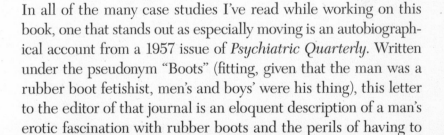

In all of the many case studies I've read while working on this book, one that stands out as especially moving is an autobiographical account from a 1957 issue of *Psychiatric Quarterly*. Written under the pseudonym "Boots" (fitting, given that the man was a rubber boot fetishist, men's and boys' were his thing), this letter to the editor of that journal is an eloquent description of a man's erotic fascination with rubber boots and the perils of having to hide this "devil of a torment" from others throughout his life.* But Boots's tale is also a celebratory ode to the transformative

---

*Consistent with sexual-imprinting theory, Boots believed that his fetish for these objects stemmed from his roughhousing with other boys: "My early childhood and schoolday pleasures became associated with the rubber boots my playmates wore. When wrestling, I would usually end up being 'the fellow underneath,' and the victor (I was secretly glad to have it so) would clamp my feet between his booted legs. Or I would deliberately push my head between a boy's boot-clad legs, and he would stanchion my head between them, like a cow is stanchioned in a dairy barn" (Boots [pseud.], "The Feelings of a Fetishist," *Psychiatric Quarterly* 10 (1957): 745).

power of human friendship. The fetishistic author realized that if
he were ever to bring his beloved boots out of the closet, as it
were, from that point on society would see him as no more than
a "weird" pervert. The prospect of forever losing his more nu-
anced, more positive social identity by letting others know of his
unusual sexual desires had long been a source of great unease to
him. Boots was quite bright, and whether or not anyone else
could see it, he knew there was a lot more to him than the fact
that his lovers were born on a production line in Boise. "It is possible for fetishists to be 'raving mad' about their fetishes," he
writes, "but outside of an irresistible, compelling obsession for
them, be in all other respects as intelligent and sane as the
president of the United States" (this was of course *years* before
George W. Bush came around to muddy that claim, but you get
Boots's point). In any event, it seems that while Boots was out
searching for used boots one day (he did so under the guise of a
hobbyist or scrap-rubber trader), he happened to make a new
friend—a "*true* friend," he stresses repeatedly. "Normal men
cannot comprehend, or fully understand, any odd, unusual feelings of this sort that are entirely foreign to their own natures . . .
However, some persons do exist who possess the rare gift of profound understanding when it comes to sensing the secret sor-
rows that shroud the lives of many folks." Boots then proceeds to
dedicate page after page to the virtues of this sympathetic new
compadre of his, an unnamed figure whom he describes as a
"normal, married heterosexual person with an independent business of his own":

> This friend is one who fully fits the best description of a
> true friend that I have ever come across: "A true friend is
> one who knows ALL ABOUT US, but still remains our
> friend." With a sense of humility, knowing that each man
> has a weakness of some kind, he accepted the truth of my
> strange and haunting obsession . . . he knew I was in many

other ways not too greatly different from many other men. My friend did not add to the weight of "my secret cross" by shunning me. He provides my ailment with palliatives to ease my "fetishistic hunger" whenever necessary.

And by this, Boots means that not only was his "true friend" a nice guy and a good listener but he even went out of his way to supply him with the objects he desired most:

Knowing that these cast-off articles of footwear are a peculiar treasure of mine, he collects all he can obtain to give to me as sentimental keepsakes that symbolize my tragic and unreturned love of man for man . . . [He] does not condemn, ridicule or scorn. My friend is not a psychiatrist, but he has done more to contribute to my happiness and peace of mind than any psychiatrist trying to chase elusive bats out of the belfry possibly could. It is a "peaceful co-existence" between two persons whose sexual emotions are as different as night is from day.

Charles Dickens wasn't a homosexual boot fetishist (at least as far as we know), but he did pen that immemorial line, "It was the best of times, it was the worst of times." I think the same can be said for where we are today as a society when it comes to sex and sexuality. In fact, we're not altogether unlike the primmer characters in *A Tale of Two Cities*, many of whom discover that their ethical sensibilities and cherished traditions are being strained or upended altogether by the French Revolution. Our own "Enlightenment" is the unprecedented pace at which the science of human sexuality is now advancing, and with it that gathering storm of data showing that deviance is more status quo than any of us imagined. Like the French monarchy, our sexual morality was built on the unsteady grounds of myths and customs, which can clearly no longer sustain it under the deluge of facts now raining down. There's no question that we're presently standing at a moral crossroads and getting soaking wet from anger and confusion in the process; the question, rather, is where to go from here. And we could use a nice solid pair of boots of our own these days to keep from slipping as we step forward. (By the way, if you want to spend some private time with them out in the woods first, have at it, I say.)

Now, if we go in the direction of those still mourning the loss of the "good old days," which, as we've seen, weren't so good at all for the erotic outliers among us, then we'd be continuing to harm others with our closed minds, fostering their "personal distress," while heading knowingly into the terminal sunset of an outdated worldview. Going in the other direction may seem obvious enough, but it's a far rockier road than it appears at first glance, which is precisely why we've been idling before it for so long. It's not just the road less traveled; it's the road never traveled. Since no society has ever ventured this way before, doing so requires us to lay brand-new tracks for a sturdier framework of sexual ethics and morality, one that those following in our footsteps can trust never to give way beneath them or fall crashing down upon their fragile heads.

To avoid the type of decay that's rotted away the entirety of that other man-made path, our new value system would need to be constructed of the brick and mortar of established scientific facts, its bedrock being the incontrovertible truth that sexual orientations are never chosen. It must also have walls of iron to protect us from the howling winds sure to arrive as we move along, walls forged by the knowledge that there is no evil but that which comes from thinking there is so. To guide us forward, we must emblazon every star in the sky with the reminder that a lustful thought is not an immoral act. And our handrails would have to be painstakingly carved from the logic that in the absence of demonstrable harm the inherent subjectivity of sex makes it a matter of private governance. Finally, and most imposing of all, we'd each have to promise to walk this brave new path completely naked from here to eternity, removing this weighty plumage of sexual normalcy and strutting, proudly, our more deviant sexual selves.

You go first.

NOTES

vii   *"Rarely has man"*: Alfred C. Kinsey, Wardell B. Pomeroy, Clyde E. Martin, and Paul H. Gebhard, *Concepts of Normality and Abnormality in Sexual Behavior* (New York: Grune and Stratton, 1949), 16.

1. WE'RE ALL PERVERTS

7   *In his 1956 play*: Jean Genet, *The Balcony* (1956; New York: Grove Press, 1994).

7   *"When it's over"*: Ibid., 35.

8   *"I consider nothing"*: Publius Terentius Afer [Terence], "Heauton Timorumenos" [The Self-Tormentor], in *Comoediae: Andria, Heauton Timorumenos, Eunuchus, Phormio, Hecyra, Adelphoe*, ed. Robert Kauer and Wallace M. Lindsay (Oxford: Oxford University Press, 1926), 25.

9   *the British lexicographer*: Thomas Blount, *Glossographia; or, A Dictionary Interpreting the Hard Words of Whatsoever Language Now in Our Refined English Tongue with Etymologies, Definitions, and Historical Observations on the Same* (1656; Ann Arbor, Mich.: EEBO Editions, ProQuest, 2010).

9   *an earlier form*: Boethius, *The Consolation of Philosophy*, rev. ed., trans. Victor Watts (524; London: Penguin Classics, 2000).

10   *Dawkins encourages his fellow*: Richard Dawkins, *The God Delusion* (London: Great Bantam Press, 2008).

10   *I've penned my own*: Jesse Bering, *The Belief Instinct: The Psychology of Souls, Destiny, and the Meaning of Life* (New York: W. W. Norton, 2011).

10   *But of all these*: Jon Jureidini, "Perversion: Erotic Form of Hatred or Exciting Avoidance of Reality?," *Journal of the American Academy of Psychoanalysis and Dynamic Psychiatry* 29 (2001): 195–211.

11 *in the work of the Victorian-era scholar*: Havelock Ellis and John A. Symonds, *Sexual Inversion* (London: Wilson Macmillan, 1897).

12 *chided the woman*: Phyllis Grosskurth, *Havelock Ellis: A Biography* (New York: Knopf, 1980).

13 *"It was never to me vulgar"*: Havelock Ellis, *My Life: The Autobiography of Havelock Ellis* (Cambridge, Mass.: Houghton Mifflin, 1930), 84.

13 *"[It's] not extremely uncommon"*: Ibid.

14 *"As there was between David"*: Richard Ellmann, *Oscar Wilde* (New York: First Vintage Books, 1988), 463.

15 *In a scathing review*: William Noyes, "Das Konträre Geschlechtsgefühl," *Psychological Review* 4, no. 4 (1897): 447.

17 *"In this way, which"*: Mervin Glasser, "Identification and Its Vicissitudes as Observed in the Perversions," *International Journal of Psychoanalysis* 67 (1986): 14.

19 *citing other nonmonogamous species*: Christopher Ryan and Cecilda Jethá, *Sex at Dawn: The Prehistoric Origins of Modern Sexuality* (New York: Harper, 2010).

20 *As the primatologist*: Frans de Waal, "Sociosexual Behavior Used for Tension Regulation in All Age and Sex Combinations Among Bonobos," *Pedophilia: Biosocial Dimensions* (1990): 378–93.

22 *"mere breath on the air"*: Jean-Paul Sartre, *No Exit, and Three Other Plays* (1948; New York: Random House, 1989), 43.

22 *"white bear effect"*: Daniel Wegner, *White Bears and Other Unwanted Thoughts: Suppression, Obsession, and the Psychology of Mental Control* (New York: Viking, 1989).

24 *negative stereotypes can*: Robert Kurzban and Mark Leary, "Evolutionary Origins of Stigmatization: The Functions of Social Exclusion," *Psychological Bulletin* 127, no. 2 (2001): 187–208.

25 *half of all "farm-bred"*: Alfred C. Kinsey, Wardell B. Pomeroy, and Clyde E. Martin, *Sexual Behavior in the Human Male* (Philadelphia: W. B. Saunders, 1948).

25 *Just as it's impossible*: Christopher M. Earls and Martin L. Lalumière, "A Case Study of Preferential Bestiality," *Archives of Sexual Behavior* 38, no. 4 (2009).

26 *But the hysteria over*: Robert F. Oaks, "'Things Fearful to Name': Sodomy and Buggery in Seventeenth-Century New England," *Journal of Social History* 12, no. 2 (1978).

27 *"butt one eye for use"*: Ibid., 275.

27 *"a faire & white skinne & head"*: Ibid., 276.

27 *"Immedyatly there appeared a working of lust"*: Ibid.

28 *highlighted by the case*: Edward Payson Evans, *The Criminal Prosecution and Capital Punishment of Animals* (1906; Clark, N.J.: Lawbook Exchange, 2009).

29 *many people of European*: Fernando L. Mendez, Joseph C. Watkins, and Michael F. Hammer, "Neandertal Origin of Genetic Variation at the

Cluster of OAS Immunity Genes," *Molecular Biology and Evolution,* published online January 12, 2013.

31 *"In all the criminal law"*: Kinsey, Pomeroy, Martin, and Gebhard, *Concepts of Normality and Abnormality in Sexual Behavior,* 12.

31 *Back in 2001*: Jonathan Haidt, "The Emotional Dog and Its Rational Tail: A Social Intuitionist Approach to Moral Judgment," *Psychological Review* 108, no. 4 (2001): 814–34.

32 *"A man belongs to"*: Roberto Gutierrez and Roger Giner-Sorolla, "Anger, Disgust, and Presumption of Harm as Reactions to Taboo-Breaking Behaviors," *Emotion* 7, no. 4 (2007): 868.

33 *"My brother is my boyfriend"*: Thomas Rodgers, "Gay Porn's Most Shocking Taboo," *Salon,* www.salon.com/2010/05/21/twincest/.

2. DAMN DIRTY APES

35 *"The butting of his haunches"*: D. H. Lawrence, *Lady Chatterley's Lover* (1928; New York: Penguin, 2006), 171.

36 *"In a small but not"*: Havelock Ellis, *Studies in the Psychology of Sex* (online-ebooks.info, 2004), 5:12.

37 *put the idea*: A. James Giannini et al., "Sexualization of the Female Foot as a Response to Sexually Transmitted Epidemics: A Preliminary Study," *Psychological Reports* 83, no. 2 (1998): 491–98.

38 *region can play host*: Roy Levin, "Smells and Tastes—Their Putative Influence on Sexual Activity in Humans," *Sexual and Relationship Therapy* 19, no. 4 (2004): 451–62.

39 *DNA sequencing reveals*: Guillaume Bourque, Pavel A. Pevzner, and Glenn Tesler, "Reconstructing the Genomic Architecture of Ancestral Mammals: Lessons from Human, Mouse, and Rat Genomes," *Genome Research* 14, no. 4 (2004): 507–16.

39 *If you take a healthy*: Paul C. Koch and Roger H. Peters, "Suppression of Adult Copulatory Behaviors Following LiCl-Induced Aversive Contingencies in Juvenile Male Rats," *Developmental Psychobiology* 20 (1987): 603–11.

40 *Sex is so corporeal*: Jamie L. Goldenberg, Tom Pyszczynski, Jeff Greenberg, and Sheldon Solomon, "Fleeing the Body: A Terror Management Perspective on the Problem of Human Corporeality," *Personality and Social Psychology Review* 4, no. 3 (2000): 200–218.

41 *"our libido thrives on"*: Sigmund Freud, *On the Universal Tendency to Debasement in the Sphere of Love* (1912; London: Hogarth Press, 1957), 187.

41 *"in its strength enjoys"*: Sigmund Freud, *Three Essays on the Theory of Sexuality,* Standard Edition (1905; London: Hogarth Press, 1953), 152.

41 *willing to taste*: Levin, "Smells and Tastes."

42 *The psychologists behind*: Richard J. Stevenson, Trevor I. Case, and Megan J. Oaten, "Effect of Self-Reported Sexual Arousal on Responses to

Sex-Related and Non-sex-related Disgust Cues," *Archives of Sexual Behavior* 40, no. 1 (2011): 79–85.

44   *"If your personal identity":* Free Republic, www.fstdt.com/QuoteComment.aspx?QID=88214&Page=3.

44   *"Homosexuals should never":* Free Republic, http://209.157.64.200/focus/f-chat/2917142/posts.

45   *"Just look at these guys!":* Ernst Hiemer, *Der Giftpilz: Ein Stürmerbuch für Jung und Alt.* (Nürnberg, 1938).

45   *how the trick:* Yoel Inbar, David Pizarro, and Paul Bloom, "Disgusting Smells Cause Decreased Liking of Gay Men," *Emotion* 12, no. 1 (2012): 23–27.

47   *express an "affectionate distaste":* Levin, "Smells and Tastes."

47   *"Girl Scent is a new":* Ibid., 454.

47   *"There does not appear to be":* Ibid., 458.

48   *When these MHC:* Claus Wedekind, Thomas Seebeck, Florence Bettens, and Alexander J. Paepke, "MHC-Dependent Mate Preferences in Humans," *Proceedings of the Royal Society of London: Series B: Biological Sciences* 260, no. 1359 (1995): 245–49.

50   *The overall effect:* Ilene L. Bernstein and Mary M. Webster, "Learned Taste Aversions in Humans," *Physiology and Behavior* 25, no. 3 (1980): 363–66.

50   *unhappily gay male:* Barry M. Maletzky and Frederick S. George, "The Treatment of Homosexuality by 'Assisted' Covert Sensitization," *Behaviour Research and Therapy* 11, no. 4 (1973): 655–57.

52   *layers of caveat:* Robert Trivers, *Social Evolution* (Menlo Park, Calif.: Benjamin/Cummins, 1985).

54   *"need to 'lower the guard'":* Stevenson, Case, and Oaten, "Effect of Self-Reported Sexual Arousal on Responses to Sex-Related and Non-sex-related Disgust Cues," 84.

55   *rate the likelihood:* Hart Blanton and Meg Gerrard, "Effect of Sexual Motivation on Men's Risk Perception for Sexually Transmitted Disease: There Must Be 50 Ways to Justify a Lover," *Health Psychology* 16, no. 4 (1997): 374.

57   *"His body erect":* Georges Bataille, *Story of the Eye,* trans. Joachim Neugroschel (1928; San Francisco: City Lights Books, 1987), 77.

57   *"Now that his balls":* Ibid., 78.

57   *Nearly a century:* Dan Ariely and George Loewenstein, "The Heat of the Moment: The Effect of Sexual Arousal on Sexual Decision Making," *Journal of Behavioral Decision Making* 19, no. 2 (2006): 87–98.

59   *"a missing tooth":* Henry Miller, *Tropic of Cancer* (1934; New Orleans: Quaint Press Books, 2012), 125.

59   *study by the anthropologist:* Daniel M. T. Fessler and C. David Navarrete, "Domain-Specific Variation in Disgust Sensitivity Across the Menstrual Cycle," *Evolution and Human Behavior* 24, no. 6 (2003): 406–17.

61 *Seventy percent of*: Nichole Fairbrother and S. Rachman, "Feelings of
    Mental Pollution Subsequent to Sexual Assault," *Behaviour Research
    and Therapy* 42, no. 2 (2004): 173–89.

61 *two groups of female*: Nichole Fairbrother, Sarah J. Newth, and
    S. Rachman, "Mental Pollution: Feelings of Dirtiness Without Physi-
    cal Contact," *Behaviour Research and Therapy* 43, no. 1 (2005):
    121–30.

62 *"I could stand and stare"*: G. B. Rahm, B. Renck, and K. C. Ringsberg,
    "'Disgust, Disgust Beyond Description'—Shame Cues to Detect Shame
    in Disguise, in Interviews with Women Who Were Sexually Abused
    During Childhood," *Journal of Psychiatric and Mental Health Nursing*
    13, no. 1 (2006): 105.

63 *measurable physical distance*: John B. Pryor, Glenn D. Reeder, Christo-
    pher Yeadon, and Matthew Hesson-McInnis, "A Dual-Process Model of
    Reactions to Perceived Stigma," *Journal of Personality and Social Psy-
    chology* 87, no. 4 (2004): 436.

63 *Suicide rates skyrocket*: Shanta R. Dube et al., "Long-Term Conse-
    quences of Childhood Sexual Abuse by Gender of Victim," *American
    Journal of Preventive Medicine* 28, no. 5 (2005): 430–38.

63 *A study by*: George A. Bonanno et al., "When the Face Reveals What
    Words Do Not: Facial Expressions of Emotion, Smiling, and the Willing-
    ness to Disclose Childhood Sexual Abuse," *Journal of Personality and
    Social Psychology* 83, no. 1 (2002): 94–110.

65 *"By the age of"*: Gilbert Herdt and Martha McClintock, "The Magical
    Age of 10," *Archives of Sexual Behavior* 29, no. 6 (2000): 596.

65 *case of a Pennsylvania man*: *Times Leader*, http://archives.timesleader
    .com/2012_5/2012_02_08_Man_who_tainted_co_workers_rsquo_yogurt
    _gets_2_years_in_prison_-news.html.

3. SISTER NYMPH AND BROTHER SATYR

67 *"People in general"*: Ogai Mori, *Vita Sexualis*, trans. Sanford Goldstein
    (1946; North Clarendon, Vt.: Tuttle, 2001), 151.

68 *"It's the abundance"*: German E. Berrios, "Classic Text No. 66: 'Madness
    from the Womb,'" *History of Psychiatry* 17, no. 2 (2006): 231.

69 *"From the same seminal"*: Ibid.

69 *"openly before all the world"*: Ibid., 232.

69 *"The genital parts"*: Ibid., 234.

70 *The feminist sociologist*: Carol Groneman, "Nymphomania: The Histori-
    cal Construction of Female Sexuality," *Signs* 19, no. 2 (1994): 337–67.

70 *a twenty-six-year-old doctor*: Ibid.

71 *"If she continued"*: Ibid., 338.

72 *the most infamous*: Ibid.

73 *"While I was praying"*: Ibid., 358.

73 *"as soon as I"*: Ibid., 357.

73 *an improbable name*: John Studd and Anneliese Schwenkhagen, "The Historical Response to Female Sexuality," *Maturitas* 63, no. 2 (2009): 107–11.

74 *She "found herself"*: Albert Ellis and Edward Sagarin, *Nymphomania: A Study of the Oversexed Woman* (New York: Gramercy, 1964), 61.

74 *"She wanted the whole"*: Ibid., 62.

75 *"Several homosexual-nymphomaniacal"*: Ibid., 74.

75 *"I told Gail"*: Ibid., 78.

76 *"If she wanted to try"*: Ibid.

77 *"He was doing the right"*: Ibid., 80.

77 *"Boys, are you guilty"*: John Harvey Kellogg, *Plain Facts for Old and Young* (1887; Project Gutenberg, 2006), www.gutenberg.org/files/19924 /19924-h/19924-h.htm.

78 *"[Circumcision] should be performed"*: Ibid.

78 *"Such a victim"*: Ibid.

78 *"the scourge of the human"*: Jeffrey Jensen Arnett, "G. Stanley Hall's *Adolescence*: Brilliance and Nonsense," *History of Psychology* 9, no. 3 (2006): 192.

79 *Fast-forward to modern*: Global Health Europe, http://globalhealth europe.org.

79 *There's one notable*: Richard von Krafft-Ebing, *Psychopathia Sexualis: With Special Reference to the Antipathic Sexual Instinct* (Google eBook, 1886).

81 *"he could no longer"*: Ibid., 52.

81 *"He said that he often suffered"*: Ibid., 51.

83 *For a very long time*: Rory C. Reid, Bruce N. Carpenter, and Thad Q. Lloyd, "Assessing Psychological Symptom Patterns of Patients Seeking Help for Hypersexual Behavior," *Sexual and Relationship Therapy* 24, no. 1 (2009): 47–63.

83 *"Satyrs display more"*: Franklin S. Klaf, *Satyriasis: A Study of Male Nymphomania* (New York: Lancer Books, 1966), 93.

84 *"Successful encounters led"*: Wayne A. Myers, "Addictive Sexual Behavior," *American Journal of Psychotherapy* (1995): 476.

86 *"excessive expressions"*: Martin P. Kafka, "The Paraphilia-Related Disorders: A Proposal for a Unified Classification of Nonparaphilic Hypersexuality Disorders," *Sexual Addiction and Compulsivity: The Journal of Treatment and Prevention* 8, nos. 3–4 (2001): 231.

88 *"which appropriately can"*: Charles Moser, "Hypersexual Disorder: Just More Muddled Thinking," *Archives of Sexual Behavior* 40, no. 2 (2011): 228.

88 *"Cultivating whatever"*: Giacomo Casanova, *History of My Life* (New York: Alfred A. Knopf, 1977), 20.

88 *Some of the best*: Kinsey, Pomeroy, and Martin, *Sexual Behavior in the Human Male*.

90 *advent of the Internet*: Al Cooper, "Sexuality and the Internet: Surfing into the New Millennium," *CyberPsychology and Behavior* 1, no. 2 (1998): 187–93.

90 *a set of statistics*: United Families International, http://unitedfamiliesinternational.wordpress.com/2010/06/02/14-shocking-pornography-statistics/.

91 *"The work of the penis"*: Barry S. Hewlett and Bonnie L. Hewlett, "Sex and Searching for Children Among Aka Foragers and Ngandu Farmers of Central Africa," *African Study Monographs* 31, no. 3 (2010): 112.

92 *Unmarried women*: Alfred C. Kinsey, Wardell B. Pomeroy, Clyde E. Martin, and Paul H. Gebhard, *Sexual Behavior in the Human Female* (Bloomington: Indiana University Press, 1953).

92 *In a 2006 survey*: Niklas Långström and R. Karl Hanson, "High Rates of Sexual Behavior in the General Population: Correlates and Predictors," *Archives of Sexual Behavior* 35, no. 1 (2006): 37–52.

95 *"She expired on"*: Subramanian Senthilkumaran et al., "Hypersexuality in a 28-Year-Old Woman with Rabies," *Archives of Sexual Behavior* 40, no. 6 (2011): 1327–28.

95 *condition has likely*: Ralph Lilly, Jeffrey L. Cummings, D. Frank Benson, and Michael Frankel, "The Human Klüver-Bucy Syndrome," *Neurology* 33, no. 9 (1983): 1141–45.

95 *Since then, Klüver-Bucy syndrome*: Julie Devinsky, Oliver Sacks, and Orrin Devinsky, "Klüver–Bucy Syndrome, Hypersexuality, and the Law," *Neurocase* 16, no. 2 (2010): 140–45.

96 *"the overlying skin"*: Krafft-Ebing, *Psychopathia Sexualis*, 51.

4. CUPID THE PSYCHOPATH

99 *"Winged Cupid, rash"*: Lucius Apuleius, *The Golden Ass*, trans. William Adlington, http://ancienthistory.about.com/library/bl/bl_cupidandpsyche.htm.

100 *"Who flies with"*: Ibid.

100 *"I pray thee"*: Ibid.

101 *"Variatio delectat!"*: Wilhelm Stekel, *Sexual Aberrations: The Phenomena of Fetishism in Relation to Sex*, trans. S. Parker (1930; New York: Liveright, 1971), 169.

102 *"pretty young girl"*: Ibid., 143.

102 *"The sight, image"*: Ibid., 145.

102 *We know they date*: Ibid.

105 *a male phenomenon*: James M. Cantor, Ray Blanchard, and Howard E. Barbaree, "Sexual Disorders," in *Oxford Textbook of Psychopathology*, 2nd ed., ed. Paul H. Blaney and Theodore Millon (New York: Oxford University Press, 2009), 527–48.

107 *newborn male rat*: Thomas J. Fillion and Elliot M. Blass, "Infantile Experience with Suckling Odors Determines Adult Sexual Behavior in Male Rats," *Science* 231, no. 4739 (1986): 729–31.

108 *goats were raised*: Keith Kendrick et al., "Sex Differences in the Influence of Mothers on the Sociosexual Preferences of Their Offspring," *Hormones and Behavior* 40, no. 2 (2001): 322–38.

109 *a chimp raised*: Maurice Temerlin, *Lucy: Growing Up Human* (Palo Alto, Calif.: Science and Behavior Books, 1975).

113 *The psychologist James Cantor*: Cantor, Blanchard, and Barbaree, "Sexual Disorders."

114 *Freund argued*: Kurt Freund, "Courtship Disorder: Is This Hypothesis Valid?," *Annals of the New York Academy of Sciences* 528, no. 1 (1988): 172–82.

114 *"phase of eye-talk"*: John Money, "Paraphilias: Phenomenology and Classification," *American Journal of Psychotherapy* 38, no. 2 (1984): 174.

115 *"push its way onto"*: Ibid.

116 *purchase clothes*: Jennifer J. Argo, Darren W. Dahl, and Andrea C. Morales, "Positive Consumer Contagion: Responses to Attractive Others in a Retail Context," *Journal of Marketing Research* 45, no. 6 (2008): 690–701.

117 *The psychotherapist*: Amy Marsh, "Love Among the Objectum Sexuals," *Electronic Journal of Human Sexuality* 13 (2010), www.ejhs.org/volume13/ObjSexuals.htm.

118 *for a documentary*: BBC, *Married to the Eiffel Tower* (Blink Films, 2008), www.youtube.com/watch?v=POTA5aZxbQA.

118 *"Well, Libby is always"*: Marsh, "Love Among the Objectum Sexuals."

118 *"I'm kind of a heavyset"*: Ibid.

119 *"I press a thousand"*: Arthur Scherr, "Voltaire's 'Candide': A Tale of Women's Equality," *Midwest Quarterly* 34, no. 3 (1993), www.accessmylibrary.com/coms2/summary_0286-142613_ITM.

119 *"Whenever a lovely"*: Stekel, *Sexual Aberrations*, 171.

119 *"He liked especially"*: Ibid., 170.

120 *"No power in the world"*: Ibid.

120 *"possess[ing] all of the"*: Ibid., 171.

121 *a socially awkward*: Brian E. McGuire, George L. Choon, Parva Nayer, and Julia Sanders, "An Unusual Paraphilia: Case Report of Oral Partialism," *Sexual and Marital Therapy* 13, no. 2 (1998): 207–10.

121 *"The patient's primary"*: Ibid., 208.

123 *"Once a man's sexual"*: Roy F. Baumeister, "Gender Differences in Erotic Plasticity: The Female Sex Drive as Socially Flexible and Responsive," *Psychological Bulletin* 126, no. 3 (2000): 348.

123 *Chivers has found*: Meredith L. Chivers, "A Brief Review and Discussion of Sex Differences in the Specificity of Sexual Arousal," *Sexual and Relationship Therapy* 20, no. 4 (2005): 377–90.

125 *"The costs of non-responding"*: Samantha J. Dawson, Kelly D. Suschinsky, and Martin L. Lalumière, "Habituation of Sexual Responses in Men and Women: A Test of the Preparation Hypothesis of Women's Genital

Responses," *Journal of Sexual Medicine*, www.ncbi.nlm.nih.gov/pubmed /23320579.

127 *mother-child relationship*: Henry Massie and Nathan Szajnberg, "The Ontogeny of a Sexual Fetish from Birth to Age 30 and Memory Processes: A Research Case Report from a Prospective Longitudinal Study," *International Journal of Psychoanalysis* 78 (1997): 755–71.

128 *"The only quirk"*: Ibid., 760.

130 *"This fantasy became"*: Thomas N. Wise and Ram Chandran Kalyanam, "Amputee Fetishism and Genital Mutilation: Case Report and Literature Review," *Journal of Sex and Marital Therapy* 26, no. 4 (2000): 340.

132 *"Like a cluster"*: Money, "Paraphilias," 165.

133 *"[Paraphilias] are not"*: Ibid., 175.

133 *The Zeigarnik effect*: Robert L. Munroe and Mary Gauvain, "Why the Paraphilias? Domesticating Strange Sex," *Cross-Cultural Research* 35, no. 1 (2001): 44–64.

135 *"We no longer burn"*: Kinsey, Pomeroy, Martin, and Gebhard, *Concepts of Normality and Abnormality in Sexual Behavior*, 21.

5. IT'S SUBJECTIVE, MY DEAR

137 *"While we are dealing here"*: Pauline Réage, *Story of O* (1954; New York: Ballantine Books, 1973), xxix.

137 *Around 11 percent*: Alan W. Shindel and Charles A. Moser, "Why Are the Paraphilias Mental Disorders?," *Journal of Sexual Medicine* 8, no. 3 (2011): 927–29.

139 *Before the more recent case*: Chris Francescani, "New York City 'Cannibal Cop' Convicted of Plot to Kidnap Women," Reuters, March 12, 2013. www.reuters.com/article/2013/03/12/us-usa-crime-cannibal-idUSBRE 92B0OX20130312.

139 *"Looking for a well-built"*: Luke Harding, "Victim of Cannibal Agreed to Be Eaten," *Guardian*, December 3, 2003, www.guardian.co.uk/world /2003/dec/04/germany.lukeharding.

141 *The man was*: E. A. Gutheil, "A Rare Case of Sadomasochism," *American Journal of Psychotherapy* 1, no. 1 (1947): 88.

142 *"But this type"*: Ibid.

145 *Rind and his colleagues*: Bruce Rind, Philip Tromovitch, and Robert Bauserman, "A Meta-analytic Examination of Assumed Properties of Child Sexual Abuse Using College Samples," *Psychological Bulletin* 124, no. 1 (1998): 22–53.

147 *radio talk-show*: Laura Schlessinger, *The Dr. Laura Program*, Premiere Radio Networks, March 23, 1999.

147 *incident without precedent*: Scott O. Lilienfeld, "When Worlds Collide: Social Science, Politics, and the Rind et al. (1998) Child Sexual Abuse Meta-analysis," *American Psychologist* 57, no. 3 (2002): 176–88.

148 *researchers still questioning*: Anna Salter, *Predators: Who They Are, How They Operate, and How We Can Protect Ourselves and Our Children* (New York: Basic Books, 2003).

148 *in 2006, the psychologist*: Heather Ulrich, Mickey Randolph, and Shawn Acheson, "Child Sexual Abuse: A Replication of the Meta-analytic Examination of Childhood Sexual Abuse by Rind, Tromovitch, and Bauserman 1998," *Scientific Review of Mental Health Practice* 4, no. 2 (2005–2006): 47–51.

150 *no hard-and-fast threshold*: Sarah-Jayne Blakemore and Suparna Choudhury, "Development of the Adolescent Brain: Implications for Executive Function and Social Cognition," *Journal of Child Psychology and Psychiatry* 47, nos. 3–4 (2006): 296–312.

151 *first such age-restricting*: Stephen Robertson, "Age of Consent Laws," http://chnm.gmu.edu/cyh/teaching-modules/230.

153 *"More than 800 years"*: Ibid.

155 *"He was depressed"*: Ratnin Dewaraja, "Formicophilia, an Unusual Paraphilia, Treated with Counseling and Behavior Therapy," *American Journal of Psychotherapy* 41, no. 4 (1987): 594.

155 *"He was [now] engaging"*: Ibid., 596.

159 *around 3 percent*: Niklas Långström and J. Kenneth Zucker, "Transvestic Fetishism in the General Population," *Journal of Sex and Marital Therapy* 31, no. 2 (2005): 87–95.

160 *one big difference*: Anne A. Lawrence, "Becoming What We Love: Autogynephilic Transsexualism Conceptualized as an Expression of Romantic Love," *Perspectives in Biology and Medicine* 50, no. 4 (2007): 506–20.

162 *controversy over*: Ray Blanchard, "The Concept of Autogynephilia and the Typology of Male Gender Dysphoria," *Journal of Nervous and Mental Disease* 177, no. 10 (1989): 616–23.

162 *Blanchard didn't just*: Ray Blanchard, I. G. Racansky, and Betty W. Steiner, "Phallometric Detection of Fetishistic Arousal in Heterosexual Male Cross-Dressers," *Journal of Sex Research* 22, no. 4 (1986): 452–62.

163 *"sex life manifested"*: Anonymous, Kinsey Institute Library and Special Collections, Indiana University.

164 *sneeze fetishist from London*: Michael B. King, "Sneezing as a Fetishistic Stimulus," *Sexual and Marital Therapy* 5, no. 1 (1990): 69–72.

164 *"Much of his intense"*: Ibid., 69.

6. A SUITABLE AGE

167 *"He was about 16"*: Testimony of Oscar Wilde in His Libel Trial, http://law2.umkc.edu/faculty/projects/ftrials/wilde/Wildelibelowfact.html.

170 *Performing a fifteen-year content*: Melanie-Angela Neuilly and Kristen Zgoba, "Assessing the Possibility of a Pedophilia Panic and Contagion Effect Between France and the United States," *Victims and Offenders* 1, no. 3 (2006): 225–54.

172 *"pre-homosexual"*: J. Michael Bailey and Kenneth J. Zucker, "Childhood Sex-Typed Behavior and Sexual Orientation: A Conceptual Analysis and Quantitative Review," *Developmental Psychology* 31, no. 1 (1995): 43–55.

175 *Complicating matters even more*: Michael C. Seto, R. Karl Hanson, and Kelly M. Babchishin, "Contact Sexual Offending by Men with Online Sexual Offenses," *Sexual Abuse: A Journal of Research and Treatment* 23, no. 1 (2011): 124–45.

175 *the more controversial*: Milton Diamond, Eva Jozifkova, and Petr Weiss, "Pornography and Sex Crimes in the Czech Republic," *Archives of Sexual Behavior* 40, no. 5 (2011): 1037–43.

176 *statistics in Japan*: Milton Diamond and A. Uchiyama, "Pornography, Rape, and Sex Crimes in Japan," *International Journal of Law and Psychiatry* 22, no. 1 (1999): 1–22.

176 *and Denmark*: Berl Kutchinsky, "The Effect of Easy Availability of Pornography on the Incidence of Sex Crimes: The Danish Experience," *Journal of Social Issues* 29, no. 3 (1973): 163–81.

177 *"We do not approve"*: Diamond, Jozifkova, and Weiss, "Pornography and Sex Crimes in the Czech Republic," 1042.

177 *"When the monster"*: Michel Foucault, *Abnormal: Lectures at the Collège de France, 1974–1975* (New York: Picador, 2003), 56.

179 *Cantor's big scoop*: James M. Cantor et al., "Cerebral White Matter Deficiencies in Pedophilic Men," *Journal of Psychiatric Research* 42, no. 3 (2008): 167–83.

179 *"The 'white matter' is the shorthand"*: Cord Jefferson, "Born This Way: Sympathy and Science for Those Who Want to Have Sex with Children," *Gawker*, September 7, 2012, http://gawker.com/5941037.

180 *thirteen hundred anonymous men*: Pekka Santtil et al., "Childhood Sexual Interactions with Other Children Are Associated with Lower Preferred Age of Sexual Partners Including Sexual Interest in Children in Adulthood," *Psychiatry Research* 175, no. 1 (2010): 154–59.

180 *There are data showing*: John A. Hunter, Aurelio Jose Figueredo, and Neil M. Malamuth, "Developmental Pathways into Social and Sexual Deviance," *Journal of Family Violence* 25, no. 2 (2010): 141–48.

181 *Seto describes several*: Michael C. Seto, *Pedophilia and Sexual Offending Against Children: Theory, Assessment, and Intervention* (Washington, D.C.: American Psychological Association, 2007).

182 *less than 1 percent*: Michael Seto, personal communication, November 18, 2012.

184 *Czech military approached*: "Remembering Kurt Freund," *Association for the Treatment of Sexual Abusers Forum* 18, no. 4 (Fall 2006), http://newsmanager.commpartners.com/atsa/issues/2006-09-15/2.html.

184 *lived through the Holocaust*: Ibid.

185 *the "Walking Nudes" version of the test*: Kurt Freund and Robin J. Watson, "Assessment of the Sensitivity and Specificity of a Phallometric Test:

An Update of Phallometric Diagnosis of Pedophilia," *Psychological Assessment: A Journal of Consulting and Clinical Psychology* 3, no. 2 (1991): 254–60.

186 *crippled by excruciating pain*: "Remembering Kurt Freund."

187 *"A certain degree of relaxation"*: Kurt Freund, "A Laboratory Method for Diagnosing Predominance of Homo- or Hetero-Erotic Interest in the Male," *Behaviour Research and Therapy* 1, no. 1 (1963): 90.

187 *hefty dose of Viagra*: Nathan J. Kolla et al., "Double-Blind, Placebo-Controlled Trial of Sildenafil in Phallometric Testing," *Journal of the American Academy of Psychiatry and the Law Online* 38, no. 4 (2010): 502–11.

187 *"light of my life"*: Vladimir Nabokov, *Lolita* (1955; New York: Vintage International, 1997), 9.

187 *"Humbert was perfectly"*: Ibid., 20.

189 *not uncommon for*: Ray Blanchard et al., "Absolute Versus Relative Ascertainment of Pedophilia in Men," *Sexual Abuse: A Journal of Research and Treatment* 21, no. 4 (2009): 431–41.

190 *age of menarche*: Sarah E. Anderson and Aviva Must, "Interpreting the Continued Decline in the Average Age at Menarche: Results from Two Nationally Representative Surveys of U.S. Girls Studied 10 Years Apart," *Journal of Pediatrics* 147, no. 6 (2005): 753–60.

192 *eponymous British pediatrician*: James M. Tanner, *Foetus into Man: Physical Growth from Conception to Maturity* (Cambridge, Mass.: Harvard University Press, 1978).

192 *"Testicular volume between"*: Jay R. Feierman, "Pedophilia (Part I): Its Relationship to the Homosexualities and the Roman Catholic Church," *Antonianum* 85, no. 3 (2010): 467.

193 *90 percent of the victims*: Ibid.

193 *Plato famously claimed*: David West, *Reason and Sexuality in Western Thought* (New York: Polity, 2005).

193 *"If I were to see"*: John Money, interview, *Paidika: The Journal of Paedophilia* 2, no. 3 (1991): 5.

194 *traveling warrior societies*: Randolph Trumbach, "Homosexual Behavior and Western Culture in the 18th Century," *Journal of Social History* 11, no. 1 (1977): 1–33.

194 *Today's ultraconservative*: Ibid.

194 *Lesbians aren't without*: Judith Gay, "'Mummies and Babies' and Friends and Lovers in Lesotho," *Journal of Homosexuality* 11, nos. 3–4 (1986): 97–116.

195 *held indefinitely*: Karen Franklin, "The Public Policy Implications of 'Hebephilia': A Response to Blanchard et al. (2008)," *Archives of Sexual Behavior* 38, no. 3 (2009): 319–20.

195 *the psychiatrist*: Jerome C. Wakefield, "Evolutionary Versus Prototype Analyses of the Concept of Disorder," *Journal of Abnormal Psychology* 108, no. 3 (1999): 374–99.

196 *such as "gender identity disorder"*: Eric Cameron, "APA to Remove 'Gender Identity Disorder' from DSM-V," *Human Rights Campaign Blog*, December 4, 2012, www.hrc.org/blog/entry/apa-to-remove-gender -identity-disorder-from-dsm-5.

196 *"normal form of human"*: Barry S. Anton, "Proceedings of the American Psychological Association for the Legislative Year 2009: Minutes of the Annual Meeting of the Council of Representatives and Minutes of the Meetings of the Board of Directors," *American Psychologist* 65, no. 5 (2010): 385–475.

197 *bear fewer offspring*: Raymond Hayes and Ray Blanchard, "Anthropo- logical Data Regarding the Adaptiveness of Hebephilia," *Archives of Sexual Behavior* 41, no. 4 (2012): 745–47.

197 *the risk of cuckoldry*: Todd K. Shackelford and Aaron T. Goetz, "Adapta- tion to Sperm Competition in Humans," *Current Directions in Psycho- logical Science* 16, no. 1 (2007): 47–50.

197 *it could help to explain*: Katarina Alanko et al., "Evidence for Heritability of Adult Men's Sexual Interest in Youth Under Age 16 from a Population- Based Extended Twin Design," *Journal of Sexual Medicine* (In Press).

198 *forensic psychologist Karen Franklin*: Karen Franklin, "Why the Rush to Create Dubious New Sexual Disorders?," *Archives of Sexual Behavior* 38, no. 4 (2010): 819–20.

198 *Green titled one*: Richard Green, "Sexual Preference for 14-Year-Olds as a Mental Disorder: You Can't Be Serious!!," *Archives of Sexual Behavior* 39, no. 3 (2010): 585–86.

201 *"Little girls are hopeless"*: Thomas J. Holt, Kristie R. Blevins, and Nata- sha Burkert, "Considering the Pedophile Subculture Online," *Sexual Abuse: A Journal of Research and Treatment* 22, no. 1 (2010): 11.

205 *"deliberately manipulat[ing]"*: Brook Barnes, "A Topless Photo Threat- ens a Major Disney Franchise," *New York Times*, April 28, 2008, www .nytimes.com/2008/04/28/business/media/28hannah.html?_r=0.

205 *"I'm sorry that my portrait"*: "Annie Leibovitz: 'Miley Cyrus Photos Were Misinterpreted,'" *Hollywood.com*, April 28, 2008, www.hollywood .com/news/celebrities/5226536/annie-leibovitz-miley-cyrus-photos-were -misinterpreted.

206 *Her famous book*: Germaine Greer, *The Female Eunuch* (New York: McGraw-Hill, 1971).

206 *visual meandering*: Germaine Greer, *The Beautiful Boy* (New York: Riz- zoli, 2003).

206 *"full of pictures"*: Emma Young, "Sticks and Stones May Break Bones but Not Stereotypes," *Sydney Morning Herald*, October 27, 2003, www.smh .com.au/articles/2003/10/26/1067103267287.html?from=storyrhs.

206 *"I know that the only"*: Germaine Greer, "Country Notebook: Beautiful Boys Cause Bedlam," *Telegraph*, December 7, 2002. www.telegraph.co.uk /gardening/3307166/Country-notebook-beautiful-boys-cause-bedlam.html.

207 *"absolutely revolting"*: Brian Simpson, "Sexualizing the Child: The Strange Case of Bill Henson, His 'Absolutely Revolting' Images, and the Law of Childhood Innocence," *Sexualities* 14, no. 3 (2011): 290–311.

207 *a philistine*: Ibid.

208 *"at best inconvenient"*: Rachel Olding, "Henson: Photo Furore Was 'Inconvenient at Best,'" *Sydney Morning Herald*, March 8, 2011, www .smh.com.au/entertainment/art-and-design/henson-photo-furore-was -inconvenient-at-best-20110307-1blbh.html.

7. LIFE LESSONS FOR THE LEWD AND LASCIVIOUS

209 *"Indeed, of all"*: Yukio Mishima, *Confessions of a Mask*, trans. Meredith Weatherby (1949; New York: New Directions, 1958), 72.

209 which set sail: Thomas W. Higginson, *Life of Francis Higginson, First Minister in the Massachusetts Bay Colony* (New York: Dodd, Mead, 1891).

210 *"five beastly Sodomiticall"*: Robert F. Oaks, "'Things Fearful to Name': Sodomy and Buggery in Seventeenth-Century New England," *Journal of Social History* 12, no. 2 (1978).

211 *testicles would be replaced*: Gregorio Marañón, *The Evolution of Sex and Intersexual Conditions* (London: G. Allen & Unwin, 1932).

212 *busy grafting monkey*: Ibid.

214 *"Posture, demeanor"*: Charles Samson Féré, *Scientific and Esoteric Studies in Sexual Degeneration in Mankind and in Animals*, trans. Ulrich van der Horst (1899; New York: Falstaff Press, 1932), 145.

214 *"there have been noticed"*: Ibid.

214 *"A doctor had [dealt with]"*: Ibid.

215 *"inability to learn"*: Ibid.

215 *"The sight which [shocks]"*: Ibid., 225.

216 *"It should be remembered"*: Ibid., 146.

217 *species of primate*: Daniel J. Povinelli, Jesse M. Bering, and Steve Giambrone, "Toward a Science of Other Minds: Escaping the Argument by Analogy," *Cognitive Science* 24, no. 3 (2000): 509–41.

217 *researchers who coined*: David Premack and Guy Woodruff, "Does the Chimpanzee Have a Theory of Mind?," *Behavioral and Brain Sciences* 1, no. 4 (1978): 515–26.

220 *it usually is with science*: Kurt Gray et al., "More Than a Body: Mind Perception and the Nature of Objectification," *Journal of Personality and Social Psychology* 101, no. 6 (2011): 1207–20.

221 *"The idea that [nudity]"*: Ibid., 1209.

222 *"perpetual virginity of the soul"*: Christopher Cahill, "Second Puberty: The Later Years of W. B. Yeats Brought His Best Poetry, Along with Personal Melodrama on an Epic Scale," *Atlantic Monthly*, December 2003.

224 *"I saw him watching"*: Angela Carter, *The Bloody Chamber, and Other Stories* (1979; New York: Penguin Books, 1993), 11.

225 *"There are some eyes"*: Ibid., 86.

226 *In a 1979 study*: Donald L. Mosher and Kevin E. O'Grady, "Homosexual Threat, Negative Attitudes Toward Masturbation, Sex Guilt, and Males' Sexual and Affective Reactions to Explicit Sexual Films," *Journal of Consulting and Clinical Psychology* 47, no. 5 (1979): 860.

227 *Our brains systematically collect*: Francis T. McAndrew and Megan A. Milenkovic, "Of Tabloids and Family Secrets: The Evolutionary Psychology of Gossip," *Journal of Applied Social Psychology* 32, no. 5 (2006): 1064–82.

229 *"devil of a torment"*: Boots [pseud.], "The Feelings of a Fetishist," *Psychiatric Quarterly* 10 (1957): 742–58.

230 *"It is possible"*: Ibid., 744.

230 *"Normal men cannot"*: Ibid., 748.

230 *"This friend is one"*: Ibid.

231 *"Knowing that these"*: Ibid., 752.

231 *"It was the best of times"*: Charles Dickens, *A Tale of Two Cities* (1859; Mineola, N.Y.: Dover, 1999), 1.

ACKNOWLEDGMENTS

When I first set out to write "a lighthearted book about sexual deviance," most people thought I was crazy. And many, I'm sure, still do after reading it. But for seeing the good in this project from its very inception, I'd like to thank my agent, Peter Tallack. I'd been writing a lot about human sexuality for *Scientific American*, and I'd been mulling over the odder social realities of being gay for most of my life, but the idea for *Perv* only crystallized while I was playing tourist in Europe one sweaty summer a few years ago. Juan and I were on a public bus that had quickly become standing room only (and all the more aromatic for it). As the bus bounded down a bumpy road to Cinque Terre, I noticed a local man around sixty or so in a cream-colored nylon suit. He was leaking from every pore (wearing such an outfit on a crowded bus in the Italian Riviera in July would have that effect), and I watched him stealing cautious glances at the chest of a voluble, and quite oblivious, teenage girl in a bikini. She looked to be about fourteen, fifteen . . . it was hard to tell, but certainly not the age of female he *should* be eyeing like that. There was nothing else to it: just a perspiring Italian pensioner staring at a pretty young girl with lust in his eyes. But it got me thinking about what scientists might be missing about our species's hidden sexual minds. So that night I hammered out a quick e-mail to Peter about the possibility of writing a book on the psychology of people's unspoken (or rather, unspeakable) desires. He liked the idea, and presto . . . here's *Perv*.

Not quite "presto" of course. In fact, the distance separating an author's hazy ideas for a book and getting that book into a shape that people would want to read is incredibly vast. It's so vast, in fact, that anyone who'd take on such a project as an editor would have to be crazier than the author. Luckily, I found such a mental state in Amanda Moon at Farrar, Straus and Giroux, who has served with grace, patience, and good humor as my editor on this and other projects over the years. Sometimes I can get a bit lost in my writing, and in the spirit of the flow I'll make a few insensitive missteps here and there—okay, "missteps" is putting it too mildly, more like the type of stumbling over sentences that can break a reader's bone. Amanda understands me well enough by now to know what it is I really *meant* to say, not that dreadfulness of a line I've just vomited onto the page without considering how people could misinterpret it. In other words, it's because of Amanda that this book isn't as offensive as it would otherwise have been, and whatever you did find offensive in it would have been inordinately more so if not for her.

I was all the more fortunate to have benefited from the talent and knowledge of Amanda's editorial assistant at FSG, Christopher Richards, who took on the role—and courageously so—of a second editor on this project. Chris had a major part in the making of *Perv*, pushing me to try harder at my weaker moments and consistently offering the level of feedback and quality of insight of an editor far beyond his years. Closer to the production end, the editorial assistant Daniel Gerstle was also instrumental in briging this project to fruition, patiently handling my many questions and emails with remarkable aplomb. For their diligence, keen eyes, and attention to detail, I would like to thank my copy editor, Ingrid Sterner, as well as my production editor, Lenni Wolff. Kathy Daneman, my publicist at FSG, is a force to be reckoned with who has worked tirelessly behind the scenes for me on all my deviant projects. I'm grateful to have her.

I'm also grateful to the very kind Robert Allen and his staff at Macmillan Audio for their work on the audiobook version of *Perv*, as well as Doug Young and Madeline Toy at Transworld Publishers in the United Kingdom. The hardworking folks at *Scientific American*, particularly Mariette DiChristina and Bora Zivkovic, have been hugely supportive and encouraging of my writing on human sexuality from the get-go. And I'd like to extend my appreciation to the helpful staff at the Kinsey Institute Library & Special Collections at Indiana University for permitting me access to their archived material.

For those in my life who for so long have had to endure my tales of sexual deviance—at dinner parties, at family reunions, and during transatlantic flights—I sincerely apologize. But I really did appreciate your generous listening. I'd like to thank my dad and stepmother; also, my sister and brother and their respective mates, and along with them my nieces and nephews, who one day will come across this book and realize why their parents always seemed to avoid their questions about what Uncle Jesse did for a living.

To Juan, my accommodating partner whose ears have sighed for so long from hearing me read aloud to him bedtime stories he'd rather not go to bed knowing and into whose loving arms I've crawled each night, wounded by the sex lives of strangers, I swear I'll never write another word about sex. Well, I promise a breather, anyway.

Finally, I am grateful to the many supportive friends and colleagues who've read and commented on early versions of this material, who answered my clueless questions about their fascinating research, or who simply offered patience and an encouraging word or two when I needed it along the way. These include (in no particular order): Synnove Heggoy, Ira Berliner, Jayne Roth, Laurie Santos, Ray Blanchard, Mike Seto, James Cantor,

Todd Shackelford, Christopher Ryan, Karen Franklin, Sean Massey, Roy Baumeister, Meredith Chivers, Dave Bjorklund, Derren Brown, Anne Lawrence, David Buss, Ben Hale, Michael Bailey, David Pizzaro, Becky Fortgang, Dan Savage, Patti Virgadamo, Erin O'Donnell, Rita Marker, Martin Kafka, Steve and Sonni Roth, Cyndi and Jesper Sjögren, Symone and Maxx Carroll, Daniel Engber, Laura Helmuth, Ginger and Brent Geiger, Alex Lakeman, Eileen Curtayne and Srini Dixit, Caroline Clark Rivera, Michèle Roten Baumgartner, Ericka L'Abbe, Alondra and Yeri Quiles, and Justin Nolan. *Thank you.*

INDEX